Soil Mechanics
Fifth Edition Solutions Manual

USTRATH

Soil Mechanics
Fifth Edition
Solutions Manual

R.F. Craig

Department of Civil Engineering
University of Dundee, UK

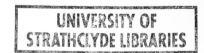
CHAPMAN & HALL
University and Professional Division

London · Glasgow · New York · Tokyo · Melbourne · Madras

Published by Chapman & Hall, 2-6 Boundary Row, London SE1 8HN

Chapman & Hall, 2-6 Boundary Row, London SE1 8HN, UK

Blackie Academic & Professional, Wester Cleddens Road,
Bishopbriggs, Glasgow G64 2NZ, UK

Chapman & Hall, 29 West 35th Street, New York NY10001, USA

Chapman & Hall Japan, Thomson Publishing Japan, Hirakawacho
Nemoto Building, 6F, 1-7-11 Hirakawa-cho, Chiyoda-ku, Tokyo 102,
Japan

Chapman & Hall Australia, Thomas Nelson Australia, 102 Dodds
Street, South Melbourne, Victoria 3205, Australia

Chapman & Hall India, R. Seshadri, 32 Second Main Road, CIT East,
Madras 600 035, India

First edition 1992
Reprinted 1993

© 1992 R.F. Craig

Typeset in 10/12pt Times by Pure Tech Corporation, India
Printed in Great Britain by Ipswich Book Co.

ISBN 0 412 47230 9   0 442 31701 8(USA)

A catalogue record for this book is available from the British Library
Library of Congress Cataloging-in-Publication Data available

Contents

Author's note

In order not to short-circuit the learning process it is vital that the reader should attempt the problems before referring to the solutions in this manual.

Basic Characteristics of Soils

1.1 British system

Soil E consists of 98% coarse material (31% gravel size; 67% sand size) and 2% fines. It is classified as SW: well-graded gravelly SAND or, in greater detail, well graded slightly silty very gravelly SAND.

Soil F consists of 63% coarse material (2% gravel size; 61% sand size) and 37% non-plastic fines (i.e. between 35% and 65% fines), therefore the soil is classified as MS: sandy SILT.

Soil G consists of 73% fine material (i.e. between 65% and 100% fines) and 27% sand size. The liquid limit is 32 and the plasticity index is 8 (i.e. $32 - 24$), plotting marginally below the A-line in the ML zone on the plasticity chart. Thus the classification is ML: SILT (M-SOIL) of low plasticity. (The plasticity chart is given in Fig. 1.6.)

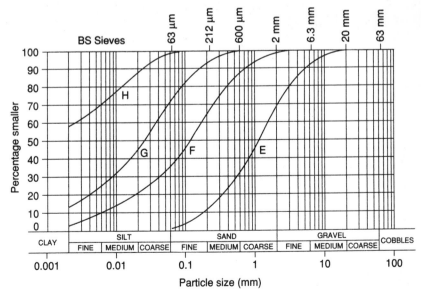

Fig Q1.1

Soil H consists of 99% fine material (58% clay size; 47% silt size). The liquid limit is 78 and the plasticity index is 47 (i.e. 78 – 31), plotting above the A-line in the CV zone on the plasticity chart. Thus the classification is CV: CLAY of very high plasticity.

Unified system

Soil E is classified as SW, a well-graded gravelly sand. More than 50% of the soil is of sand size and the fine-grained fraction is less than 5%. The following values are obtained from the particle size distribution curve:

$$D_{10} = 0.16 \, \text{mm}; \; D_{30} = 0.53 \, \text{mm}; \; D_{60} = 1.40 \, \text{mm}$$

$$\therefore \;\; C_u = \frac{1.40}{0.16} = 8.8 \qquad \text{(equation 1.1)}$$

$$C_z = \frac{0.53^2}{1.40 \times 0.16} = 1.25 \qquad \text{(equation 1.2)}$$

$$\text{i.e. } C_u > 6 \text{ and } 1 < C_z < 3$$

Soil F is classified as SM, a silty sand. The coarse-grained fraction is 63% and the fine-grained fraction 37%. Virtually all the coarse-grained fraction is of sand size. The fine-grained fraction is non-plastic.

Soil G is classified as ML, an inorganic silt with slight plasticity. The coarse-grained fraction is 27% and the fine-grained fraction 73% (60% silt size; 13% clay size). The liquid limit is 32 and the plasticity index is 8 (i.e. 32 – 24), plotting marginally below the A-line in the ML zone on the plasticity chart (Fig. 1.7).

Soil H is classified as CH, an inorganic clay of high plasticity. Virtually all the soil is fine-grained, 58% being of clay size and 41% of silt size. The liquid limit is 78 and the plasticity index is 47 (i.e. 78 – 31), plotting above the A-line in the CH zone on the plasticity chart (Fig. 1.7.)

1.2

From equation 1.17:

$$1 + e = G_s(1 + w) \frac{\rho_w}{\rho} = 2.70 \times 1.095 \times \frac{1.00}{1.91} = 1.55$$

$$\therefore e = 0.55$$

Using equation 1.13:

$$S_r = \frac{wG_s}{e} = \frac{0.095 \times 2.70}{0.55} = 0.466 \qquad (46.6\%)$$

Using equation 1.19:

$$\rho_{sat} = \frac{G_s + e}{1 + e} \rho_w = \frac{3.25}{1.55} \times 1.00 = 2.10 \, \text{Mg/m}^3$$

From equation 1.14:

$$w = \frac{e}{G_s} = \frac{0.55}{2.70} = 0.204 \quad (20.4\%)$$

1.3

Equations similar to 1.17 to 1.20 apply in the case of unit weights, thus:

$$\gamma_d = \frac{G_s}{1+e} \gamma_w = \frac{2.72}{1.70} \times 9.8 = 15.7 \, \text{kN/m}^3$$

$$\gamma_{sat} = \frac{G_s + e}{1+e} \gamma_w = \frac{3.42}{1.70} \times 9.8 = 19.7 \, \text{kN/m}^3$$

Using equation 1.21

$$\gamma' = \frac{G_s - 1}{1+e} \gamma_w = \frac{1.72}{1.70} \times 9.8 = 9.9 \, \text{kN/m}^3$$

Using equation 1.18a with $S_r = 0.75$:

$$\gamma = \frac{G_s + S_r e}{1+e} \gamma_w = \frac{3.245}{1.70} \times 9.8 = 18.7 \, \text{kN/m}^3$$

Using equation 1.13:

$$w = \frac{S_r e}{G_s} = \frac{0.75 \times 0.70}{2.72} = 0.193 \quad (19.3\%)$$

1.4

Volume of specimen $= \frac{\pi}{4} \times 38^2 \times 76 = 86\,200 \, \text{mm}^3$

$$\text{Bulk density: } (\rho) = \frac{\text{Mass}}{\text{Volume}} = \frac{168.0}{86\,200 \times 10^{-3}} = 1.95 \, \text{Mg/m}^3$$

$$\text{Water content: } (w) = \frac{168.0 - 130.5}{130.5} = 0.287 \quad (28.7\%)$$

From equation 1.17:

$$1 + e = G_s(1+w)\frac{\rho_w}{\rho} = 2.73 \times 1.287 \times \frac{1.00}{1.95} = 1.80 \therefore \quad e = 0.80$$

Using equation 1.13:

$$S_r = \frac{wG_s}{e} = \frac{0.287 \times 2.73}{0.80} = 0.98 \quad (98\%)$$

1.5

Using equation 1.24:

$$\rho_d = \frac{\rho}{1+w} = \frac{2.15}{1.12} = 1.92 \, \text{Mg/m}^3$$

From equation 1.17:

$$1 + e = G_s(1+w)\frac{\rho_w}{\rho} = 2.65 \times 1.12 \times \frac{1.00}{2.15} = 1.38 \therefore e = 0.38$$

Using equation 1.13:

$$S_r = \frac{wG_s}{e} = \frac{0.12 \times 2.65}{0.38} = 0.837 \quad (83.7\%)$$

Using equation 1.15:

$$A = \frac{e - wG_s}{1+e} = \frac{0.38 - 0.318}{1.38} = 0.045 \quad (4.5\%)$$

The zero air voids dry density is given by equation 1.25:

$$\rho_d = \frac{G_s}{1 + wG_s} = \frac{2.65}{1 + (0.135 \times 2.65)} \times 1.00 = 1.95 \, \text{Mg/m}^3$$

i.e. a dry density of 2.00 Mg/m^3 would not be possible.

1.6

Mass (g)	ρ (Mg/m^3)	w	ρ_d (Mg/m^3)	ρ_{d_0} (Mg/m^3)	ρ_{d_5} (Mg/m^3)	$\rho_{d_{10}}$ (Mg/m^3)
2010	2.010	0.128	1.782	1.990	1.890	1.791
2092	2.092	0.145	1.827	1.925	1.829	1.733
2114	2.114	0.156	1.829	1.884	1.790	1.696
2100	2.100	0.168	1.798	1.843	1.751	1.658
2055	2.055	0.192	1.724	1.765	1.676	1.588

In each case the bulk density (ρ) is equal to the mass of compacted soil divided by the volume of the mould. The corresponding value of dry density (ρ_d) is obtained from equation 1.24. The dry density/water content curve is plotted, from which:

$$w_{opt} = 15\% \quad \text{and} \quad \rho_{d_{max}} = 1.83 \, \text{Mg/m}^3$$

Equation 1.26, with A equal, in turn, to 0, 0.05 and 0.10, is used to calculate values of dry density (ρ_{d_0}, ρ_{d_5}, $\rho_{d_{10}}$ respectively) for use in plotting the air content curves. The experimental values of w have been used in these calculations; however, any series of w values within the relevant range could be used. By inspection, the value of air content at maximum dry density is 3.5%.

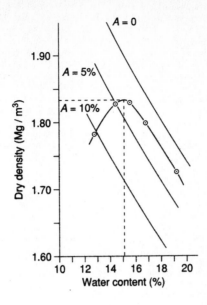

Fig Q1.6

1.7

From equation 1.20:

$$e = \frac{G_s \rho_w}{\rho_d} - 1$$

The maximum and minimum values of void ratio are given by:

$$e_{max} = \frac{G_s \rho_w}{\rho_{d_{min}}} - 1$$

$$e_{min} = \frac{G_s \rho_w}{\rho_{d_{max}}} - 1$$

From equation 1.23:

$$I_D = \frac{G_s \rho_w (1/\rho_{d_{min}} - 1/\rho_d)}{G_s \rho_w (1/\rho_{d_{min}} - 1/\rho_{d_{max}})}$$

$$= \frac{[1 - (\rho_{d_{min}}/\rho_d)] \, 1/\rho_{d_{min}}}{[1 - (\rho_{d_{min}}/\rho_{d_{max}})] \, 1/\rho_{d_{min}}}$$

$$= \left(\frac{\rho_d - \rho_{d_{min}}}{\rho_{d_{max}} - \rho_{d_{min}}} \right) \frac{\rho_{d_{max}}}{\rho_d}$$

$$= \left(\frac{1.72 - 1.54}{1.81 - 1.54} \right) \frac{1.81}{1.72}$$

$$= 0.70 \quad (70\%)$$

Seepage

2.1

The coefficient of permeability is determined from the equation:

$$k = 2.3 \frac{al}{At_1} \log \frac{h_0}{h_1}$$

$$\text{where,} \quad a = \frac{\pi}{4} \times 0.005^2 \text{ m}^2; \quad l = 0.2 \text{ m};$$

$$A = \frac{\pi}{4} \times 0.1^2 \text{ m}^2; \quad t_1 = 3 \times 60^2 \text{ s};$$

$$\log \frac{h_0}{h_1} = \log \frac{1.00}{0.35} = 0.456$$

$$\therefore \quad k = \frac{2.3 \times 0.005^2 \times 0.2 \times 0.456}{0.1^2 \times 3 \times 60^2} = 4.9 \times 10^{-8} \text{ m/s}$$

2.2

The flow net is drawn in Fig. Q2.2. In the flow net there are 3.7 flow channels and 11 equipotential drops, i.e. $N_f = 3.7$ and $N_d = 11$. The overall loss in total head is 4.00 m. The quantity of seepage is calculated by using equation 2.16:

$$q = kh \frac{N_f}{N_d} = 10^{-6} \times 4.00 \times \frac{3.7}{11} = 1.3 \times 10^{-6} \text{ m}^3/\text{s per m}$$

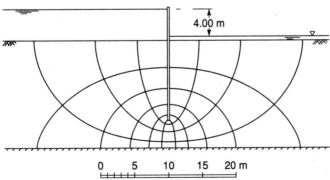

4.00 m

0   5   10   15   20 m

Fig. Q2.2

2.3

The flow net is drawn in Fig. Q2.3, from which $N_f \doteq 3.5$ and $N_d = 9$. The overall loss in total head is 3.00 m. Then:

$$q = kh\frac{N_f}{N_d} = 5 \times 10^{-5} \times 3.00 \times \frac{3.5}{9} = 5.8 \times 10^{-5} \text{ m}^3/\text{s per m}$$

The pore water pressure is determined at the points of intersection of the equipotentials with the base of the structure. The total head (h) at each point is obtained from the flow net. The elevation head (z) at each point on the base of the structure is -2.50 m. The calculations are tabulated below and the distribution of pressure (u) is plotted to scale in the figure.

Point	h (m)	$h - z$ (m)	$u = \gamma_w(h - z)$ (kN/m^2)
1	2.33	4.83	47
2	2.00	4.50	44
3	1.67	4.17	41
4	1.33	3.83	37
5	1.00	3.50	34
6	0.67	3.17	31

e.g. for Point 1:

$$h_1 = \frac{7}{9} \times 3.00 = 2.33 \text{ m}$$

$$h_1 - z_1 = 2.33 - (-2.50) = 4.83 \text{ m}$$

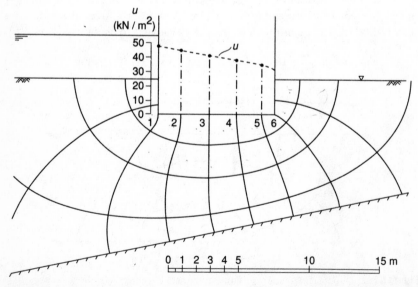

Fig. Q2.3

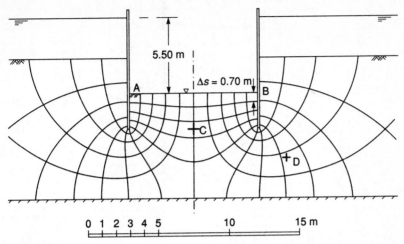

Fig. Q2.4

$$u_1 = 9.8 \times 4.83 = 47 \text{ kN/m}^2$$

The uplift force on the base of the structure is equal to the area of the pressure diagram and is 316 kN per unit length.

2.4

The flow net is drawn in Fig. Q2.4, from which $N_f = 10.0$ and $N_d = 11$. The overall loss in total head is 5.50 m. Then:

$$q = kh \frac{N_f}{N_d} = 4.0 \times 10^{-7} \times 5.50 \times \frac{10}{11} = 2.0 \times 10^{-6} \text{ m}^3/\text{s per m}$$

2.5

The flow net is drawn in Fig. Q2.5, from which $N_f = 4.2$ and $N_d = 9$. The overall loss in total head is 5.00 m. Then:

$$q = kh \frac{N_f}{N_d} = 2.0 \times 10^{-6} \times 5.00 \times \frac{4.2}{9} = 4.7 \times 10^{-6} \text{ m}^3/\text{s per m}$$

2.6

The scale transformation factor in the x-direction is given by equation 2.21, i.e.

$$x_t = x \frac{\sqrt{k_z}}{\sqrt{k_x}} = x \frac{\sqrt{1.8}}{\sqrt{5.0}} = 0.60 x$$

Thus in the transformed section the horizontal dimension 33.00 m becomes (33.00×0.60), i.e. 19.80 m, and the slope 1 : 5 becomes 1 : 3. All dimensions

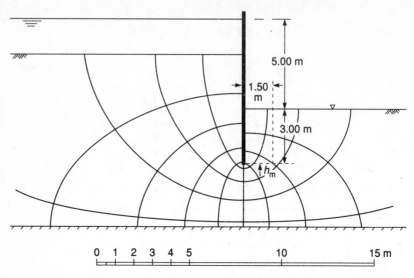

Fig. Q2.5

in the vertical direction are unchanged. The transformed section is shown in Fig. Q2.6 and the flow net is drawn as for the isotropic case. From the flow net, $N_f = 3.25$ and $N_d = 12$. The overall loss in total head is 14.00 m. The equivalent isotropic permeability applying to the transformed section is given by equation 2.23, i.e.

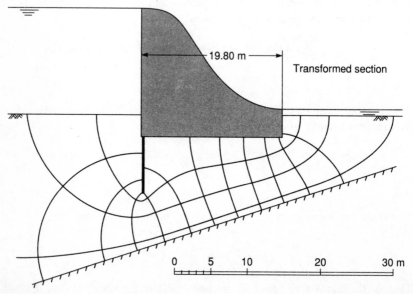

Fig. Q2.6

$$k' = \sqrt{(k_x k_z)} = \sqrt{(5.0 \times 1.8)} \times 10^{-7} = 3.0 \times 10^{-7} \text{ m/s}$$

Thus the quantity of seepage is given by:

$$q = k'h \frac{N_f}{N_d} = 3.0 \times 10^{-7} \times 14.00 \times \frac{3.25}{12} = 1.1 \times 10^{-6} \text{ m}^3/\text{s per m}$$

2.7

The scale transformation factor in the x-direction is:

$$x_t = x \frac{\sqrt{k_z}}{\sqrt{k_x}} = x \frac{\sqrt{2.7}}{\sqrt{7.5}} = 0.60 x$$

Thus all dimensions in the x-direction are multipled by 0.60. All dimensions in the z-direction are unchanged. The transformed section is shown in Fig. Q2.7. The equivalent isotropic permeability is:

$$k' = \sqrt{(k_x k_z)} = \sqrt{(7.5 \times 2.7)} \times 10^{-6} = 4.5 \times 10^{-6} \text{ m/s}$$

The focus of the basic parabola is at point A. The parabola passes through point G such that:

$$GC = 0.3HC = 0.3 \times 30 = 9.0 \text{ m}$$

Thus the co-ordinates of G are:

$$x = -48.0 \quad \text{and} \quad z = +20.0$$

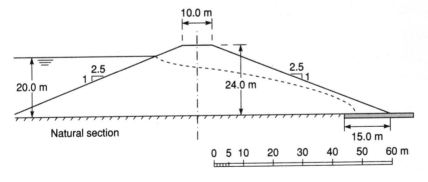

Natural section

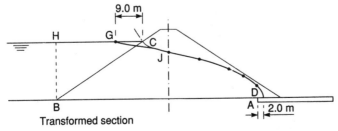

Transformed section

Fig. Q2.7

Substituting these co-ordinates in equation 2.34:

$$-48.0 = x_0 - \frac{20.0^2}{4x_0}$$

Hence:

$$x_0 = 2.0 \, \text{m}$$

Using equation 2.34, with $x_0 = 2.0 \, \text{m}$, the co-ordinates of a number of points on the basic parabola are calculated, i.e.

$$x = 2.0 - \frac{z^2}{8.0}$$

x	2.0	0	−5.0	−10.0	−20.0	−30.0
z	0	4.00	7.48	9.80	13.27	16.00

The basic parabola is plotted in Fig. Q2.7. The upstream correction is drawn, using personal judgement.

No downstream correction is required in this case since $\beta = 180°$. If required, the top flow line can be plotted back onto the natural section, the x co-ordinates above being divided by the scale transformation factor. The quantity of seepage can be calculated using equation 2.33, i.e.

$$q = 2k'x_0 = 2 \times 4.5 \times 10^{-6} \times 2.0 = 1.8 \times 10^{-5} \, \text{m}^3/\text{s per m}$$

2.8

The flow net is drawn in Fig. Q2.8, from which $N_f = 3.3$ and $N_d = 7$. The overall loss in total head is 2.8 m. Then:

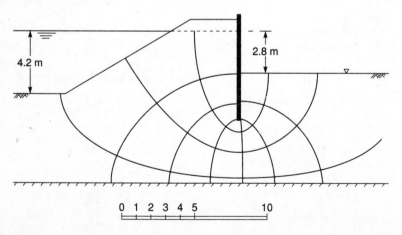

4.2 m

2.8 m

0 1 2 3 4 5   10

Fig. Q2.8

$$q = kh \frac{N_f}{N_d} = 4.5 \times 10^{-5} \times 2.8 \times \frac{3.3}{7}$$

$$= 5.9 \times 10^{-5} \text{ m}^3/\text{s per m}$$

2.9

The two isotropic soil layers, each 5 m thick, can be considered as a single homogeneous anisotropic layer of thickness 10 m in which the coefficients of permeability in the horizontal and vertical directions respectively are given by equations 2.24 and 2.25, i.e.

$$\bar{k}_x = \frac{H_1 k_1 + H_2 k_2}{H_1 + H_2} = \frac{10^{-6}}{10} \{(5 \times 2.0) + (5 \times 16)\} = 9.0 \times 10^{-6} \text{ m/s}$$

$$\bar{k}_z = (H_1 + H_2) \Big/ \left(\frac{H_1}{k_1} + \frac{H_2}{k_2} \right) = 10 \Big/ \left(\frac{5}{2 \times 10^{-6}} + \frac{5}{16 \times 10^{-6}} \right)$$

$$= 3.6 \times 10^{-6} \text{ m/s}$$

Then the scale transformation factor is given by:

$$x_t = x \frac{\sqrt{\bar{k}_z}}{\sqrt{\bar{k}_x}} = x \frac{\sqrt{3.6}}{\sqrt{9.0}} = 0.63 x$$

Thus in the transformed section the dimension 10.00 m becomes 6.30 m; vertical dimensions are unchanged. The transformed section is shown in Fig. Q2.9 and the flow net is drawn as for a single isotropic layer. From the flow net, $N_f = 5.6$ and $N_d = 11$. The overall loss in total head is 3.50 m. The equivalent isotropic permeability is:

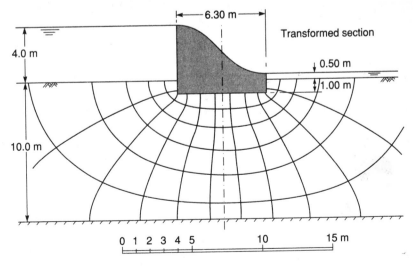

Fig. Q2.9

$$k' = \sqrt{(\bar{k_x}\bar{k_z})} = \sqrt{(9.0 \times 3.6)} \times 10^{-6} = 5.7 \times 10^{-6} \text{ m/s}$$

Then the quantity of seepage is given by:

$$q = k'h\frac{N_f}{N_d} = 5.7 \times 10^{-6} \times 3.50 \times \frac{5.6}{11}$$

$$= 1.0 \times 10^{-5} \text{ m}^3/\text{s per m}$$

Effective stress

3

3.1

Buoyant unit weight:

$$\gamma' = \gamma_{sat} - \gamma_w = 20 - 9.8 = 10.2 \, \text{kN/m}^3$$

Effective vertical stress:

$$\sigma'_v = 5 \times 10.2 = 51 \, \text{kN/m}^2 \qquad \textbf{or}$$

Total vertical stress:

$$\sigma_v = (2 \times 9.8) + (5 \times 20) = 119.6 \, \text{kN/m}^2$$

Pore water pressure:

$$u = 7 \times 9.8 = 68.6 \, \text{kN/m}^2$$

Effective vertical stress:

$$\sigma'_v = \sigma_v - u = 119.6 - 68.6 = 51 \, \text{kN/m}^2$$

3.2

Buoyant unit weight:

$$\gamma' = \gamma_{sat} - \gamma_w = 20 - 9.8 = 10.2 \, \text{kN/m}^3$$

Effective vertical stress:

$$\sigma'_v = 5 \times 10.2 = 51 \, \text{kN/m}^2 \qquad \textbf{or}$$

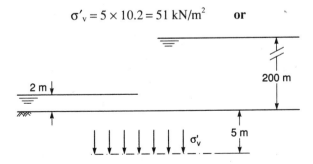

Fig. Q3.1/.2

Total vertical stress:

$$\sigma_v = (200 \times 9.8) + (5 \times 20) = 2060 \, \text{kN/m}^2$$

Pore water pressure:

$$u = 205 \times 9.8 = 2009 \, \text{kN/m}^2$$

Effective vertical stress:

$$\sigma'_v = \sigma_v - u = 2060 - 2009 = 51 \, \text{kN/m}^2$$

3.3

At top of clay:

$$\sigma_v = (2 \times 16.5) + (2 \times 19) = 71.0 \, \text{kN/m}^2$$

$$u = 2 \times 9.8 = 19.6 \, \text{kN/m}^2$$

$$\sigma'_v = \sigma_v - u = 71.0 - 19.6 = 51.4 \, \text{kN/m}^2$$

Alternatively:

$$\gamma'(\text{sand}) = 19 - 9.8 = 9.2 \, \text{kN/m}^3$$

$$\sigma'_v = (2 \times 16.5) + (2 \times 9.2) = 51.4 \, \text{kN/m}^2$$

At bottom of clay:

$$\sigma_v = (2 \times 16.5) + (2 \times 19) + (4 \times 20) = 151.0 \, \text{kN/m}^2$$

$$u = 12 \times 9.8 = 117.6 \, \text{kN/m}^2$$

$$\sigma'_v = \sigma_v - u = 151.0 - 117.6 = 33.4 \, \text{kN/m}^2$$

N.B. The alternative method of calculation is not applicable because of the artesian condition.

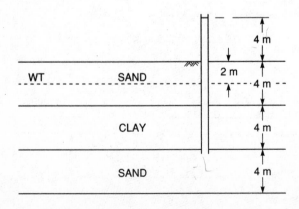

Fig. Q3.3

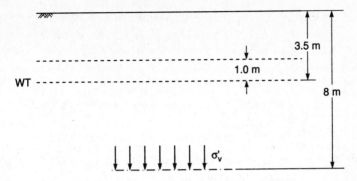

Fig. Q3.4

3.4

$$\gamma' = 20 - 9.8 = 10.2 \, \text{kN/m}^3$$

at 8 m depth:

$$\sigma'_v = (2.5 \times 16) + (1.0 \times 20) + (4.5 \times 10.2) = 105.9 \, \text{kN/m}^2$$

3.5

$$\gamma' \,(\text{sand}) = 19 - 9.8 = 9.2 \, \text{kN/m}^3$$
$$\gamma' \,(\text{clay}) = 20 - 9.8 = 10.2 \, \text{kN/m}^3$$

(a) Immediately after WT rise:

At 8 m depth, pore water pressure is governed by the new WT level because the permeability of the sand is high.

$$\therefore \quad \sigma'_v = (3 \times 16) + (5 \times 9.2) = 94.0 \, \text{kN/m}^2$$

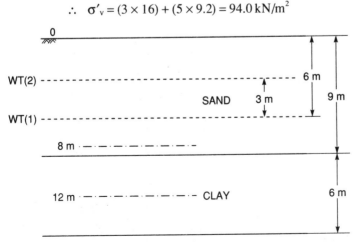

Fig. Q3.5

At 12 m depth, pore water pressure is governed by the old WT level because the permeability of the clay is very low. (However, there will be an increase in total stress of 9 kN/m² due to the increase in unit weight from 16 kN/m³ to 19 kN/m² between 3 m and 6 m depth: this is accompanied by an immediate increase of 9 kN/m² in pore water pressure.)

$$\therefore \quad \sigma'_v = (6 \times 16) + (3 \times 9.2) + (3 \times 10.2) = 154.2 \, \text{kN/m}^2$$

(b) Several years after WT rise:

At both depths, pore water pressure is governed by the new WT level, it being assumed that swelling of the clay is complete.
At 8 m depth:

$$\sigma'_v = 94.0 \, \text{kN/m}^2 \quad \text{(as above)}$$

At 12 m depth:

$$\sigma'_v = (3 \times 16) + (6 \times 9.2) + (3 \times 10.2) = 133.8 \, \text{kN/m}^2$$

3.6

Total weight:

$$\overline{ab} = 21.0 \, \text{kN}$$

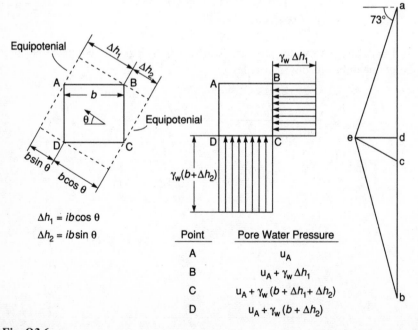

$$\Delta h_1 = ib\cos\theta$$
$$\Delta h_2 = ib\sin\theta$$

Point	Pore Water Pressure
A	u_A
B	$u_A + \gamma_w \, \Delta h_1$
C	$u_A + \gamma_w \, (b + \Delta h_1 + \Delta h_2)$
D	$u_A + \gamma_w \, (b + \Delta h_2)$

Fig. Q3.6

Effective weight:

$$\overline{ac} = 11.2 \, kN$$

Resultant boundary water force:

$$\overline{be} = 11.9 \, kN$$

Seepage force:

$$\overline{ce} = 3.4 \, kN$$

Resultant body force:

$$\overline{ae} = 9.9 \, kN \quad (73° \text{ to horizontal})$$

(Refer to Fig. Q3.6.)

3.7

Situation (1):

(a)

$$\sigma = 3\gamma_w + 2\gamma_{sat} = (3 \times 9.8) + (2 \times 20) = 69.4 \, kN/m^2$$

$$u = \gamma_w(h - z) = 9.8 \, \{1 - (-3)\} = 39.2 \, kN/m^2$$

$$\sigma' = \sigma - u = 69.4 - 39.2 = 30.2 \, kN/m^2$$

(b)

$$i = \frac{2}{4} = 0.5$$
$$j = i\gamma_w = 0.5 \times 9.8 = 4.9 \, kN/m^3 \downarrow$$
$$\sigma' = 2(\gamma' + j) = 2(10.2 + 4.9) = 30.2 \, kN/m^2$$

Situation (2):

(a)

$$\sigma = 1\gamma_w + 2\gamma_{sat} = (1 \times 9.8) + (2 \times 20) = 49.8 \, kN/m^2$$

$$u = \gamma_w(h - z) = 9.8\{1 - (-3)\} = 39.2 \, kN/m^2$$

$$\sigma' = \sigma - u = 49.8 - 39.2 = 10.6 \, kN/m^2$$

(b)

$$i = \frac{2}{4} = 0.5$$

$$j = i\gamma_w = 0.5 \times 9.8 = 4.9 \, kN/m^3 \uparrow$$

$$\sigma' = 2(\gamma' - j) = 2(10.2 - 4.9) = 10.6 \, kN/m^2$$

3.8

The flow net is drawn in Fig. Q2.4.

Loss in total head between adjacent equipotentials:

$$\Delta h = \frac{5.50}{N_d} = \frac{5.50}{11} = 0.50 \, \text{m}$$

Exit hydraulic gradient:

$$i_e = \frac{\Delta h}{\Delta s} = \frac{0.50}{0.70} = 0.71$$

The critical hydraulic gradient is given by equation 3.9:

$$i_c = \frac{\gamma'}{\gamma_w} = \frac{10.2}{9.8} = 1.04$$

Therefore, factor of safety against 'boiling' (equation 3.11):

$$F = \frac{i_c}{i_e} = \frac{1.04}{0.71} = 1.5$$

Total head at C:

$$h_C = \frac{n_d}{N_d} h = \frac{2.4}{11} \times 5.50 = 1.20 \, \text{m}$$

Elevation head at C:

$$z_C = -2.50 \, \text{m}$$

Pore water pressure at C:

$$u_C = 9.8(1.20 + 2.50) = 36 \, \text{kN/m}^2$$

Therefore, effective vertical stress at C:

$$\sigma'_C = \sigma_C - u_C = (2.5 \times 20) - 36 = 14 \, \text{kN/m}^2$$

For point D:

$$h_D = \frac{7.3}{11} \times 5.50 = 3.65 \, \text{m}$$

$$z_D = -4.50 \, \text{m}$$

$$u_D = 9.8(3.65 + 4.50) = 80 \, \text{kN/m}^2$$

$$\sigma'_D = \sigma_D - u_D = (3 \times 9.8) + (7 \times 20) - 80 = 90 \, \text{kN/m}^2$$

3.9

The flow net is drawn in Fig. Q2.5.

For a soil prism 1.50×3.00 m, adjacent to the piling:

$$h_m = \frac{2.6}{9} \times 5.00 = 1.45 \text{ m}$$

Factor of safety against 'heaving' (equation 3.10):

$$F = \frac{i_c}{i_m} = \frac{\gamma' d}{\gamma_w h_m} = \frac{9.7 \times 3.00}{9.8 \times 1.45} = 2.0$$

With a filter:

$$F = \frac{\gamma' d + w}{\gamma_w h_m}$$

$$\therefore \quad 3 = \frac{(9.7 \times 3.00) + w}{9.8 \times 1.45}$$

$$\therefore \quad w = 13.5 \text{ kN/m}^2$$

Depth of filter $= \dfrac{13.5}{21} = 0.65$ m (if above water level).

Shear strength

4

4.1

$$\sigma = 295 \, \text{kN/m}^2$$

$$u = 120 \, \text{kN/m}^2$$

$$\sigma' = \sigma - u = 295 - 120 = 175 \, \text{kN/m}^2$$

$$\tau_f = c' + \sigma' \tan \varphi' = 12 + 175 \tan 30° = 113 \, \text{kN/m}^2$$

4.2

σ'_3 (kN/m²)	$\sigma_1 - \sigma_3$ (kN/m²)	σ'_1 (kN/m²)
100	452	552
200	908	1108
400	1810	2210
800	3624	4424

The Mohr circles are drawn in Fig. Q4.2, together with the failure envelope from which $\varphi' = 44°$.

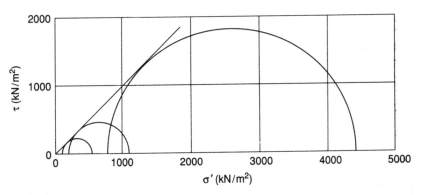

Fig. Q4.2

Alternatively, $\frac{1}{2}(\sigma_1 - \sigma_3)$ is plotted against $\frac{1}{2}(\sigma'_1 + \sigma'_3)$ and the modified failure envelope is drawn through the four points from which $\alpha' = 34\frac{3}{4}°$. Hence:

$$\varphi' = \sin^{-1}(\tan 34\frac{3}{4}°) = 44°$$

4.3

σ_3 (kN/m²)	$\sigma_1 - \sigma_3$ (kN/m²)	σ_1 (kN/m²)
200	222	422
400	218	618
600	220	820

The Mohr circles and failure envelope are drawn in Fig. Q4.3, from which $c_u = 110 \, \text{kN/m}^2$ and $\varphi_u = 0$.

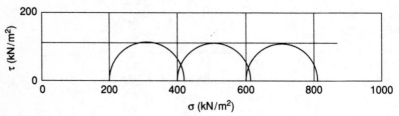

Fig. Q4.3

4.4

The modified shear strength parameters are:

$$\alpha' = \tan^{-1}(\sin \varphi') = \tan^{-1}(\sin 29°) = 26°$$

$$a' = c' \cos \varphi' = 15 \cos 29° = 13 \, \text{kN/m}^2$$

The co-ordinates of the stress point representing failure conditions in the test are:

$$\frac{1}{2}(\sigma_1 - \sigma_3) = \frac{1}{2} \times 134 = 67 \, \text{kN/m}^2$$

$$\frac{1}{2}(\sigma_1 + \sigma_3) = \frac{1}{2}(384 + 250) = 317 \, \text{kN/m}^2$$

The pore water pressure at failure is given by the horizontal distance between this stress point and the modified failure envelope. Thus from Fig. Q4.4:

$$u_f = 205 \, \text{kN/m}^2$$

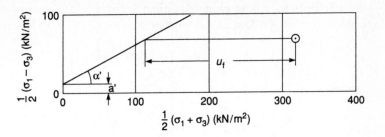

Fig. Q4.4

4.5

σ_3 (kN/m^2)	$\sigma_1 - \sigma_3$ (kN/m^2)	σ_1 (kN/m^2)	u (kN/m^2)	σ'_3 (kN/m^2)	σ'_1 (kN/m^2)
150	103	253	82	68	171
300	202	502	169	131	333
450	305	755	252	198	503
600	410	1010	331	269	679

The Mohr circles and failure envelope are drawn in Fig. Q4.5, from which $c' = 0$ and $\varphi' = 25\frac{1}{2}°$.

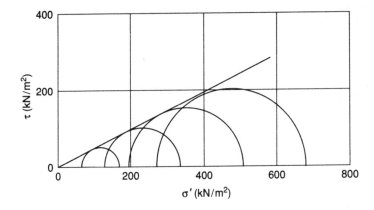

Fig. Q4.5

The principal stress difference at failure depends only on the value of all-round pressure under which consolidation took place, i.e. 250 kN/m^2. Hence, by proportion, the expected value of $(\sigma_1 - \sigma_3)_f = 170 \, kN/m^2$.

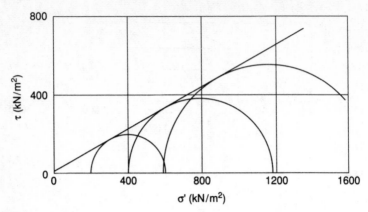

Fig. Q4.6

4.6

σ'_3 (kN/m²)	$\Delta V/V_0$	$\Delta l/l_0$	Area (mm²)	Load (N)	$\sigma_1 - \sigma_3$ (kN/m²)	σ'_1 (kN/m²)
200	0.061	0.095	1177	480	408	608
400	0.086	0.110	1165	895	768	1168
600	0.108	0.124	1155	1300	1126	1726

The average cross-sectional area of each specimen is obtained from equation 4.12; the original values of A, l and V are: $A_0 = 1134$ mm², $l_0 = 76$ mm, $V_0 = 86\,200$ mm³. The Mohr circles and failure envelope are drawn in Fig. Q4.6, from which $c' = 15$ kN/m² and $\varphi' = 28°$.

4.7

The torque required to produce shear failure is given by:

$$T = \pi dh\, c_u \frac{d}{2} + 2\int_0^{d/2} 2\pi r\, dr\, c_u r$$

$$= \pi c_u \frac{d^2 h}{2} + 4\pi c_u \int_0^{d/2} r^2\, dr$$

$$= \pi c_u \left(\frac{d^2 h}{2} + \frac{d^3}{6} \right)$$

Then:

$$35 = \pi\, c_u \left(\frac{5^2 \times 10}{2} + \frac{5^3}{6} \right) \times 10^{-3}$$

$$\therefore \quad c_u = 76 \text{ kN/m}^2$$

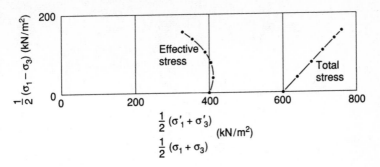

Fig. Q4.8

4.8

The relevant stress values are calculated as follows:

$$\sigma_3 = 600 \text{ kN/m}^2$$

$\sigma_1 - \sigma_3$	0	80	158	214	279	319
σ_1	600	680	758	814	879	919
u	200	229	277	318	388	433
σ'_1	400	451	481	496	491	486
σ'_3	400	371	323	282	212	167
$\frac{1}{2}(\sigma_1 - \sigma_3)$	0	40	79	107	139	159
$\frac{1}{2}(\sigma'_1 + \sigma'_3)$	400	411	402	389	351	326
$\frac{1}{2}(\sigma_1 + \sigma_3)$	600	640	679	707	739	759

The stress paths are plotted in Fig. Q4.8. The initial points on the effective and total stress paths are separated by the value of the back pressure ($u_s = 200$ kN/m²).

$$A_f = \frac{433 - 200}{319} = 0.73$$

4.9

$$B = \frac{\Delta u_3}{\Delta \sigma_3} = \frac{144}{350 - 200} = 0.96$$

ε_a (%)	$\Delta\sigma_1 = \sigma_1 - \sigma_3$ (kN/m²)	Δu_1 (kN/m²)	$\bar{A} = \dfrac{\Delta u_1}{\Delta \sigma_1}$
0	0	0	—
2	201	100	0.50
4	252	96	0.38
6	275	78	0.28
8	282	68	0.24
10	283	65	0.23

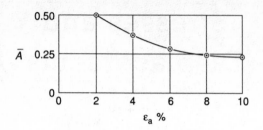

Fig. Q4.9

The variation of $\overline{A}$ with axial strain is plotted in Fig. Q4.9. At failure $\overline{A} = 0.23$.

Stresses and Displacements

5.1

Vertical stress is given by:

$$\sigma_z = \frac{Q}{z^2} I_p = \frac{5000}{5^2} I_p$$

Values of I_p are obtained from Table 5.1.

r (m)	r/z	I_p	σ_z (kN/m^2)
0	0	0.478	96
1	0.2	0.433	87
2	0.4	0.329	66
3	0.6	0.221	44
4	0.8	0.139	28
5	1.0	0.084	17
7	1.4	0.032	6
10	2.0	0.009	2

The variation of σ_z with radial distance (r) is plotted in Fig. Q5.1.

5.2

Below the centre load (Fig. Q5.2):

$$\frac{r}{z} = 0 \text{ for the 7500-kN load} \quad \therefore \quad I_p = 0.478$$

$$\frac{r}{z} = \frac{5}{4} = 1.25 \text{ for the 10 000- and 9000-kN loads}$$

$$\therefore \quad I_p = 0.045$$

Then:

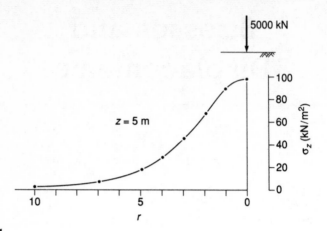

Fig. Q5.1

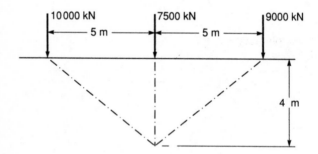

Fig. Q5.2

$$\sigma_z = \sum \left(\frac{Q}{z^2} I_p \right)$$

$$= \frac{7500 \times 0.478}{4^2} + \frac{10\,000 \times 0.045}{4^2} + \frac{9000 \times 0.045}{4^2}$$

$$= 224 + 28 + 25 = 277 \text{ kN/m}^2$$

5.3

The vertical stress under a corner of a rectangular area is given by:

$$\sigma_z = qI_r$$

where values of I_r are obtained from Fig. 5.10. In this case:

$$\sigma_z = 4 \times 250 \times I_r \ (\text{kN/m}^2)$$

$$m = n = \frac{1}{z}$$

z (m)	m, n	I_r	σ_z (kN/m^2)
0	—	—	(250)
0.5	2.00	0.233	233
1	1.00	0.176	176
1.5	0.67	0.122	122
2	0.50	0.085	85
3	0.33	0.045	45←
4	0.25	0.027	27
7	0.14	0.009	9
10	0.10	0.005	5

σ_z is plotted against z in Fig. Q5.3.

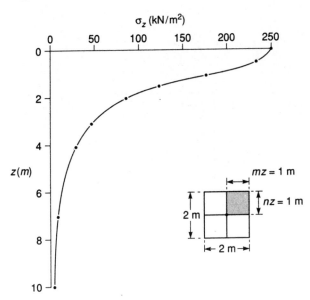

Fig. Q5.3

5.4

(a)

$$m = \frac{12.5}{12} = 1.04$$

$$n = \frac{18}{12} = 1.50$$

From Fig. 5.10, $I_r = 0.196$.

$$\therefore \quad \sigma_z = 2 \times 175 \times 0.196 = 68 \text{ kN/m}^2$$

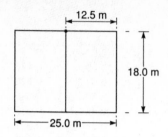

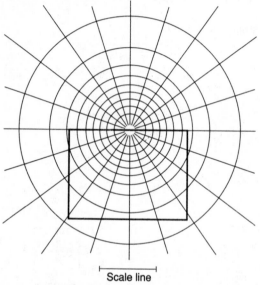

Scale line

Influence value per unit pressure = 0.005

Fig. Q5.4

(b) The foundation is drawn on Newmark's chart as shown in Fig. Q5.4, the scale line representing 12 m (z). Number of influence areas $(N) = 78$.

$$\therefore \quad \sigma_z = 0.005\,Nq = 0.005 \times 78 \times 175 = 68 \text{ kN/m}^2$$

5.5

$Q = 150 \text{ kN/m}$; $h = 4.00 \text{ m}$; $m = 0.5$. The total thrust is given by equation 5.18:

$$P_x = \frac{2Q}{\pi}\,\frac{1}{m^2 + 1} = \frac{2 \times 150}{\pi \times 1.25} = 76 \text{ kN/m}$$

Equation 5.17 is used to obtain the pressure distribution:

$$p_x = \frac{4Q}{\pi h} \frac{m^2 n}{(m^2 + n^2)^2} = \frac{150}{\pi} \frac{m^2 n}{(m^2 + n^2)^2} \text{ kN/m}^2$$

n	$\dfrac{m^2 n}{(m^2 + n^2)^2}$	p_x (kN/m^2)
0	0	0
0.1	0.370	17.7
0.2	0.595	28.4
0.3	0.649	31.0
0.4	0.595	28.4
0.6	0.403	19.2
0.8	0.252	12.0
1.0	0.160	7.6

The pressure distribution is plotted in Fig. Q5.5.

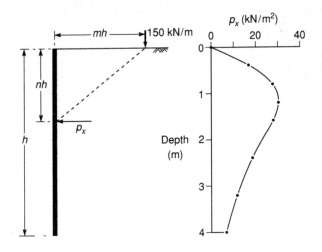

Fig. Q5.5

5.6

$$H/B = \frac{10}{2} = 5$$

$$L/B = \frac{4}{2} = 2$$

$$D/B = \frac{1}{2} = 0.5$$

Hence from Fig. 5.15:

$$\mu_1 = 0.82$$

$$\mu_0 = 0.94$$

The immediate settlement is given by equation 5.28:

$$s_i = \mu_0 \, \mu_1 \, \frac{qB}{E_u}$$

$$= 0.94 \times 0.82 \times \frac{200 \times 2}{45} = 7 \text{ mm}$$

Lateral earth pressure $\boxed{6}$

6.1

For $\varphi' = 37°$ the active pressure coefficient is given by:

$$K_a = \frac{1 - \sin 37°}{1 + \sin 37°} = 0.25 \quad \text{(or from Table 6.3)}$$

The total active thrust (equation 6.6 with $c' = 0$) is:

$$P_a = \tfrac{1}{2} K_a \gamma H^2 = \frac{1}{2} \times 0.25 \times 17 \times 6^2 = 76.5 \text{ kN/m}$$

If the wall is prevented from yielding, the at-rest condition applies. The approximate value of the coefficient of earth pressure at-rest is given by equation 6.15a:

$$K_0 = 1 - \sin \varphi' = 1 - \sin 37° = 0.40$$

and the thrust on the wall is:

$$P_0 = \tfrac{1}{2} K_0 \gamma H^2 = \frac{1}{2} \times 0.40 \times 17 \times 6^2 = 122 \text{ kN/m}$$

6.2

The active pressure coefficients for the three soil types are as follows:

$$K_{a_1} = \frac{1 - \sin 35°}{1 + \sin 35°} = 0.271$$

$$K_{a_2} = \frac{1 - \sin 27°}{1 + \sin 27°} = 0.375 \qquad \sqrt{K_{a_2}} = 0.613$$

$$K_{a_3} = \frac{1 - \sin 42°}{1 + \sin 42°} = 0.198$$

Distribution of active pressure (plotted in Fig. Q6.2):

Depth (m)	Soil	Active pressure (kN/m^2)	
3	1	$0.271 \times 16 \times 3$	$= 13.0$
5	1	$(0.271 \times 16 \times 3) + (0.271 \times 9.2 \times 2)$	$= 13.0 + 5.0 = 18.0$
5	2	$\{(16 \times 3) + (9.2 \times 2)\}\, 0.375 - (2 \times 17 \times 0.613) = 24.9 - 20.9 = 4.0$	
8	2	$4.0 + (0.375 \times 10.2 \times 3)$	$= 4.0 + 11.5 = 15.5$
8	3	$\{(16 \times 3) + (9.2 \times 2) + (10.2 \times 3)\}\, 0.198$	$= 19.2$
12	3	$19.2 + (0.198 \times 11.2 \times 4)$	$= 19.2 + 8.9 = 28.1$

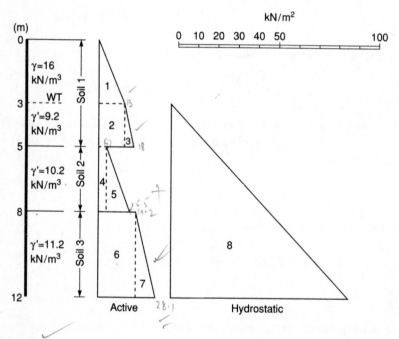

Fig. Q6.2

At a depth of 12 m, the hydrostatic pressure $= 9.8 \times 9 = 88.2\,\text{kN/m}^2$. Calculation of total thrust and its point of application (Forces are numbered as in Fig. Q6.2 and moments are taken about the top of the wall) per m:

Total thrust = 571 kN/m

Point of application is $\left(\dfrac{4893}{571}\right)$ m from the top of the wall i.e. 8.57 m

Force (kN)		Arm (m)	Moment (kNm)
(1) $\frac{1}{2} \times 0.271 \times 16 \times 3^2$	= 19.5	2.0	39.0
(2) $0.271 \times 16 \times 3 \times 2$	= 26.0	4.0	104.0
(3) $\frac{1}{2} \times 0.271 \times 9.2 \times 2^2$	= 5.0	4.33	21.7
(4) $[0.375\{(16 \times 3) + (9.2 \times 2)\}$ $\quad - \{2 \times 17 \times 0.613\}]\,3$	= 12.2	6.5	79.3
(5) $\frac{1}{2} \times 0.375 \times 10.2 \times 3^2$	= 17.2	7.0	120.4
(6) $[0.198\{(16 \times 3) + (9.2 \times 2) + (10.2 \times 3)\}]4$	= 76.8	10.0	768.0
(7) $\frac{1}{2} \times 0.198 \times 11.2 \times 4^2$	= 17.7	10.67	188.9
(8) $\frac{1}{2} \times 9.8 \times 9^2$	= 396.9	9.0	3572.1
	571.3		4893.4

6.3

(a) For $\varphi_u = 0$, $K_a = K_p = 1$

$$K_{ac} = K_{pc} = 2\sqrt{1.5} = 2.45 \qquad k_{ac} = 2\sqrt{\left(1 + \frac{c_w}{c_u}\right)}$$

At the lower end of the piling:

$$p_a = K_a q + K_a \gamma_{sat} z - K_{ac} c_u$$

$$= (1 \times 18 \times 3) + (1 \times 20 \times 4) - (2.45 \times 50)$$

$$= 54 + 80 - 122.5$$

$$= 11.5 \text{ kN/m}^2$$

$$p_p = K_p \gamma_{sat} z + K_{pc} c_u$$

$$= (1 \times 20 \times 4) + (2.45 \times 50)$$

$$= 80 + 122.5$$

$$= 202 \text{ kN/m}^2$$

(b) For $\varphi' = 26°$ and $\delta = \frac{1}{2}\varphi'$

$$K_a = 0.35 \qquad \text{(Table 6.3)}$$

$$K_{ac} = 2\sqrt{(0.35 \times 1.5)} = 1.45 \qquad \text{(Equation 6.19)} \quad k_{ac} = 2\sqrt{k_a\left(1 + \frac{c_w}{c}\right)}$$

$$K_p = 3.7 \qquad \text{(Table 6.4)}$$

$$K_{pc} = 2\sqrt{(3.7 \times 1.5)} = 4.7 \qquad \text{(Equation 6.24)} \quad K_{pc} = 2\sqrt{\left[k_p\left(1 + \frac{c_w}{c}\right)\right]}$$

At the lower end of the piling:

$$p_a = K_a q + K_a \gamma' z - K_{ac} c'$$

$$= (0.35 \times 18 \times 3) + (0.35 \times 10.2 \times 4) - (1.45 \times 10)$$

$$= 18.9 + 14.3 - 14.5$$

$$= 18.7 \, \text{kN/m}^2$$

$$p_p = K_p \gamma' z + K_{pc} c'$$

$$= (3.7 \times 10.2 \times 4) + (4.7 \times 10)$$

$$= 151 + 47$$

$$= 198 \, \text{kN/m}^2$$

6.4

For $\varphi' = 38°$, $K_a = 0.24$

$$\gamma' = 20 - 9.8 = 10.2 \, \text{kN/m}^2$$

The pressure distribution is shown in Fig. Q6.4. Consider moments (per m length of wall) about X:

Force (kN)		Arm (m)	Moment (kNm)
(1) $\frac{1}{2} \times 0.24 \times 17 \times 3.9^2$	= 31.0	4.00	124.0
(2) $0.24 \times 17 \times 3.9 \times 2.7$	= 43.0	1.35	58.0
(3) $\frac{1}{2} \times 0.24 \times 10.2 \times 2.7^2$	= 8.9	0.90	8.0
(4) $\frac{1}{2} \times 9.8 \times 2.7^2$	= 35.7	0.90	32.1
	$R_h = \underline{118.6}$		
(5) $6.2 \times 0.4 \times 23.5$	= 58.3	2.80	163.2
(6) $4.0 \times 0.4 \times 23.5$	= 37.6	2.00	75.2
(7) $3.9 \times 2.6 \times 17$	= 172.4	1.30	224.1
(8) $2.3 \times 2.6 \times 20$	= 119.6	1.30	155.5
(9) 100	= $\underline{100.0}$	2.80	$\underline{280.0}$
	$R_v = \underline{487.9}$		$\underline{1120.1} = M$

Lever arm of base resultant:

$$\frac{M}{R_v} = \frac{1120}{488} = 2.29 \, \text{m}$$

Eccentricity of base resultant:

$$e = 2.29 - 2.00 = 0.29 \, \text{m}$$

Base pressures (equation 6.25):

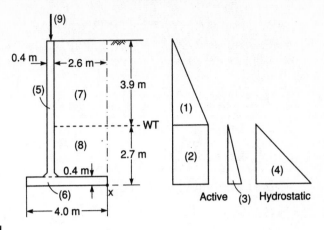

Fig. Q6.4

$$p = \frac{R_v}{B}\left(1 \pm \frac{6e}{B}\right)$$

$$= \frac{488}{4}(1 \pm 0.435)$$

$$= 175 \text{ kN/m}^2 \text{ and } 69 \text{ kN/m}^2 \quad \checkmark \quad \text{max/min}$$

Factor of safety against sliding (equation 6.26):

$$F = \frac{R_v \tan \delta}{R_h} = \frac{488 \times \tan 25°}{118} = 1.9$$

6.5

For $\varphi' = 36°$, $K_a = 0.26$ and $K_p = 3.85$

$$\frac{K_p}{F} = \frac{3.85}{2}$$

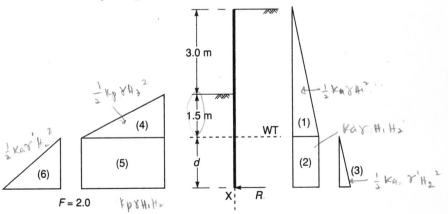

$F = 2.0$

Fig. Q6.5

$$\gamma' = 20 - 9.8 = 10.2 \text{ kN/m}^3$$

The pressure distribution is shown in Fig. Q6.5: hydrostatic pressure on the two sides of the wall balances. Consider moments about X (per m), assuming $d > 0$:

Force (kN)		Arm (m)	Moment (kNm)
(1) $\frac{1}{2} \times 0.26 \times 17 \times 4.5^2$	$= 44.8$	$d + 1.5$	$44.8d + 67.2$
(2) $0.26 \times 17 \times 4.5 \times d$	$= 19.9d$	$d/2$	$9.95d^2$
(3) $\frac{1}{2} \times 0.26 \times 10.2 \times d^2$	$= 1.33d^2$	$d/3$	$0.44d^3$
(4) $-\frac{1}{2} \times \frac{3.85}{2} \times 17 \times 1.5^2 = 36.8$		$d + 0.5$	$-36.8d - 18.4$
(5) $-\frac{3.85}{2} \times 17 \times 1.5 \times d = 49.1d$		$d/2$	$-24.55d^2$
(6) $-\frac{1}{2} \times \frac{3.85}{2} \times 10.2 \times d^2 = 9.82d^2$		$d/3$	$-3.27d^3$

$$\Sigma M = -2.83d^3 - 14.6d^2 + 8.0d + 48.8 = 0$$

$$\therefore \qquad d^3 + 5.16d^2 - 2.83d - 17.24 = 0$$

$$\therefore \qquad d = 1.79 \text{ m}$$

Depth of penetration $= 1.2(1.79 + 1.50) = 3.95$ m

$$\Sigma F = 0, \text{ hence } R = 71.5 \text{ kN (substituting } d = 1.79 \text{ m)}$$

Over additional 20% embedded depth:

$$p_p - p_a = (3.85 \times 17 \times 4.5) - (0.26 \times 17 \times 1.5) + (3.85 - 0.26)(10.2 \times 2.12)$$

$$= 365.5 \text{ kN/m}^2$$

Net passive resistance $= 365.5 \times 0.66 = 241$ kN ($> R$)

6.6

The active pressure coefficient is given by equation 6.17, in which: $\alpha = 105°$, $\beta = 20°$, $\varphi = 36°$, $\delta = 25°$

$$K_a = \left[\frac{\sin 69°/\sin 105°}{\sqrt{\sin 130°} + \dfrac{\sqrt{(\sin 61° \sin 16°)}}{\sqrt{\sin 85°}}} \right]^2 = 0.50$$

The total active thrust (acting at 25° above the normal) is given by equation 6.16:

$$P_a = \frac{1}{2} \times 0.50 \times 19 \times 7.50^2 = 267 \text{ kN/m}$$

Horizontal component:

$$P_h = 267 \cos 40° = 205 \text{ kN/m}$$

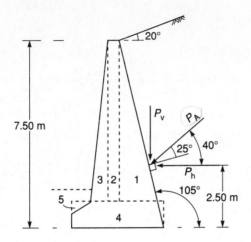

Fig. Q6.6

Vertical component:

$$P_v = 267 \sin 40° = 172 \, \text{kN/m}$$

Consider moments about the heel of the wall (Fig. Q6.6) (per m):

Force (kN)		Arm (m)	Moment (kNm)
(1) $\frac{1}{2} \times 1.75 \times 6.50 \times 23.5$	$= 133.7$	1.42	190
(2) $0.50 \times 6.50 \times 23.5$	$= 76.4$	2.25	172
(3) $\frac{1}{2} \times 0.70 \times 6.50 \times 23.5$	$= 53.5$	2.73	146
(4) $1.00 \times 4.00 \times 23.5$	$= 94.0$	2.00	188
(5) $-\frac{1}{2} \times 0.80 \times 0.50 \times 23.5$	$= -4.7$	3.73	-18
$P_a \sin 40°$	$= 172.0$	0.67	115
	$R_v = \overline{525}$		
$P_a \cos 40°$	$(R_h) = \overline{205}$	2.50	512
			$\overline{1305} = M$

Lever arm of base resultant:

$$\frac{M}{R_v} = \frac{1305}{525} = 2.49 \, \text{m}$$

Eccentricity of base resultant:

$$e = 2.49 - 2.00 = 0.49 \, \text{m}$$

Base pressures (equation 6.25):

$$p = \frac{525}{4} \left(1 \pm \frac{6 \times 0.49}{4} \right)$$

$$= 228 \, \text{kN/m}^2 \text{ and } 35 \, \text{kN/m}^2$$

Factor of safety against sliding:

(a) No passive resistance:

$$F = \frac{R_v \tan \delta}{R_h} = \frac{525 \times \tan 25°}{205} = 1.2$$

(b) Passive resistance over 1.5 m depth ($K_p = 3.85$ for $\varphi = 36°$):

$$P_p = \tfrac{1}{2} \times 3.85 \times 19 \times 1.5^2 = 82 \text{ kN/m}$$

$$F = \frac{R_v \tan \delta + P_p}{R_h} = \frac{525 \tan 25° + 82}{205} = 1.6$$

6.7

For $\varphi' = 35°$, $K_a = 0.27$

For $\varphi' = 27°$, $K_a = 0.375$ and $K_p = 2.67$

For soil, $\gamma' = 11.2 \text{ kN/m}^3$
For backfill, $\gamma' = 10.2 \text{ kN/m}^3$

The pressure distribution is shown in Fig. Q6.7. Hydrostatic pressure is balanced. Consider moments about the anchor point (A), per m.

Force (kN)			Arm (m)	Moment (kNm)
(1) $\tfrac{1}{2} \times 0.27 \times 17 \times 5^2$	=	57.4	1.83	105.0
(2) $0.27 \times 17 \times 5 \times 3$	=	68.9	5.00	344.5
(3) $\tfrac{1}{2} \times 0.27 \times 10.2 \times 3^2$	=	12.4	5.50	68.2
(4) $0.375 \{(17 \times 5) + (10.2 \times 3)\}d$	=	$43.4\,d$	$d/2 + 6.50$	$21.7d^2 + 282.1d$
(5) $\tfrac{1}{2} \times 0.375 \times 11.2 \times d^2$	=	$2.1\,d^2$	$2d/3 + 6.50$	$1.4d^3 + 13.7d^2$
(6) $-2 \times 10 \times \sqrt{0.375} \times d$	=	$-12.2\,d$	$d/2 + 6.50$	$-6.1d^2 - 79.3d$
(7) $-\tfrac{1}{2} \times \dfrac{2.67}{2} \times 11.2 \times d^2$	=	$-7.5\,d^2$	$2d/3 + 6.50$	$-5.0d^3 - 48.8d^2$
(8) $-2 \times 10 \times \dfrac{\sqrt{2.67}}{2} \times d$	=	$-16.3\,d$	$d/2 + 6.50$	$-8.2d^2 - 106.0d$
Tie rod force per m	=	$-T$	0	0

$$\Sigma M = -3.6d^3 - 27.7d^2 + 96.8d + 517.7 = 0$$

$$\therefore \quad d^3 + 7.7d^2 - 26.9d - 143.8 = 0$$

$$\therefore \quad d = 4.67 \text{ m}$$

Depth of penetration $= 1.2d = 5.60$ m

Algebraic sum of forces for $d = 4.67$ m

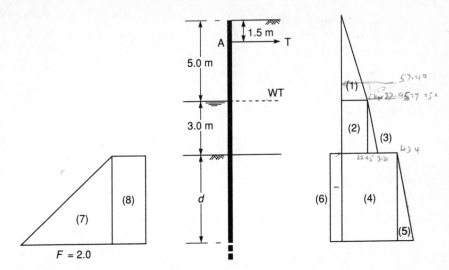

Fig. Q6.7

$$\Sigma F = 57.4 + 68.9 + 12.4 + 202.7 + 45.8 - 57.0 - 163.5 - 76.1 - T = 0$$

$$\therefore \quad T = 90.5 \text{ kN/m}$$

Force in each tie rod = $2.5 \, T = 226$ kN

6.8

For $\varphi' = 36°$, $K_a = 0.26$ and $K_p = 3.85$

$$\gamma' = 21 - 9.8 = 11.2 \text{ kN/m}^3$$

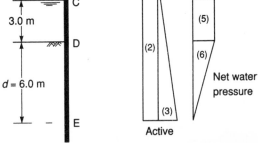

Fig. Q6.8

To allow for dredging take:

$$d = \frac{7.20}{1.2} = 6.00 \text{ m}$$

The pressure distribution is shown in Fig. Q6.8. In this case the net water pressure at C (taking $d = 6.00$ m) is given by:

$$u_C = \frac{15.0}{16.5} \times 1.5 \times 9.8 = 13.4 \text{ kN/m}^2$$

Consider moments about the anchor point A (per m)

Force (kN)		Arm (m)	Moment (kNm)
(1) $\frac{1}{2} \times 0.26 \times 18 \times 4.5^2$	= 47.4	1.5	71.1
(2) $0.26 \times 18 \times 4.5 \times 10.5$	= 221.1	8.25	1824.0
(3) $\frac{1}{2} \times 0.26 \times 11.2 \times 10.5^2$	= 160.5	10.0	1605.0
(4) $\frac{1}{2} \times 13.4 \times 1.5$	= 10.1	4.0	40.4
(5) 13.4×3.0	= 40.2	6.0	241.2
(6) $\frac{1}{2} \times 13.4 \times 6.0$	= 40.2	9.5	381.9
	519.3		4163.6
(7) $- P_{P_m}$		11.5	$- 11.5 P_{P_m}$

$$\Sigma M = 0 \quad \therefore \quad P_{P_m} = \frac{4163.6}{11.5} = 362 \text{ kN/m}$$

Available passive resistance:

$$P_p = \frac{1}{2} \times 3.85 \times 11.2 \times 6^2 = 776 \text{ kN/m}$$

Factor of safety:

$$F_p = \frac{P_p}{P_{P_m}} = \frac{776}{362} = 2.1$$

Force in each tie $= 2T = 2(519 - 362) = 314 \text{ kN}$

6.9

For $\varphi' = 30°$ and $\delta = 15°$, $K_a = 0.30$ and $K_p = 4.8$

$$\gamma' = 20 - 9.8 = 10.2 \text{ kN/m}^3$$

The pressure distribution is shown in Fig. Q6.9. Assuming uniform loss in total head along the wall, the net water pressure at C is:

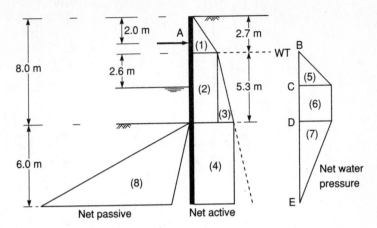

Fig. Q6.9

$$u_C = \frac{14.7}{17.3} \times 2.6 \times 9.8 = 21.6 \, \text{kN/m}^2$$

and the average seepage pressure around the wall is:

$$\bar{j} = \frac{2.6}{17.3} \times 9.8 = 1.5 \, \text{kN/m}^3$$

Consider moments about the prop (A) (per m).

Force (kN)		Arm (m)	Moment (kNm)
(1) $\frac{1}{2} \times 0.3 \times 17 \times 2.7^2$	= 18.6	− 0.20	− 3.7
(2) $0.3 \times 17 \times 2.7 \times 5.3$	= 73.0	3.35	244.5
(3) $\frac{1}{2} \times 0.3(10.2 + 1.5) \, 5.3^2$	= 49.3	4.23	208.5
(4) $0.3\{(17 \times 2.7) + (11.7 \times 5.3)\}6.0$	= 194.2	9.00	1747.8
(5) $\frac{1}{2} \times 21.6 \times 2.6$	= 28.1	2.43	68.4
(6) 21.6×2.7	= 58.3	4.65	271.2
(7) $\frac{1}{2} \times 21.6 \times 6.0$	= 64.8	8.00	518.4
			3055
(8) $\frac{1}{2}\{4.8(10.2 - 1.5) - 0.3(10.2 + 1.5)\}6.0^2 = 688.5$		10.00	6885

Factor of safety:

$$F_r = \frac{6885}{3055} = 2.25$$

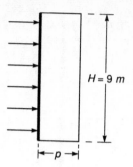

Fig. Q6.10

6.10

For $\varphi' = 40°$, $K_a = 0.22$

The pressure distribution is shown in Fig. Q6.10.

$$p = 0.65 K_a \gamma H = 0.65 \times 0.22 \times 19 \times 9 = 24.5 \text{ kN/m}^2$$

Strut load $= 24.5 \times 1.5 \times 3 = 110 \text{ kN}$

6.11

$$\gamma = 18 \text{ kN/m}^3; \quad \varphi' = 34°$$

$$H = 3.50 \text{ m}; \quad nH = 3.35 \text{ m}; \quad mH = 1.85 \text{ m}$$

Consider a trial value of $F = 2.0$. Refer to Fig. 6.33.

$$\varphi'_m = \tan^{-1}\left(\frac{\tan 34°}{2.0}\right) = 18.6°$$

Then:

$$\alpha = 45° + \frac{\varphi'_m}{2} = 54.3°$$

$$W = \tfrac{1}{2} \times 18 \times 3.50^2 \times \cot 54.3° = 79.2 \text{ kN/m}$$

$$P = \tfrac{1}{2} \times \gamma_s \times 3.35^2 = 5.61\gamma_s \text{ kN/m}$$

$$U = \tfrac{1}{2} \times 9.8 \times 1.85^2 \times \operatorname{cosec} 54.3° = 20.6 \text{ kN/m}$$

Equations 6.28 and 6.29 then become:

$$5.61\gamma_s + (N - 20.6)\tan 18.6° \cos 54.3° - N \sin 54.3° = 0$$

$$79.2 - (N - 20.6)\tan 18.6° \sin 54.3° - N \cos 54.3° = 0$$

i.e.

$$5.61\gamma_s - 0.616N - 4.05 = 0$$

$$79.2 - 0.857N + 5.63 = 0$$

$$\therefore \quad N = \frac{84.8}{0.857} = 98.9 \text{ kN/m}$$

Then:

$$5.61\gamma_s - 60.9 - 4.05 = 0$$

$$\therefore \quad \gamma_s = \frac{64.9}{5.61} = 11.6 \text{ kN/m}^3$$

The calculations for trial values of F of 2.0, 1.5 and 1.0 are summarized below:

F	φ'_m	α	W (kN/m)	U (kN/m)	N (kN/m)	γ_s (kN/m^3)
2.0	18.6°	54.3°	79.2	20.6	98.9	11.6
1.5	24.2°	57.1°	71.3	19.9	85.6	9.9
1.0	34°	62°	58.6	19.1	65.7	7.7

γ_s is plotted against F in Fig. Q6.11.

From Fig. Q6.11, for $\gamma_s = 10.6 \text{ kN/m}^3$, $F = 1.7$

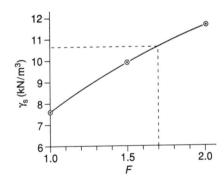

Fig. Q6.11

6.12

$$\gamma z = 18 \times 3.6 = 64 \text{ kN/m}^2$$

$$\sigma_{max} = 18 \times 3.6 \{1 + 0.30(3.6^2/5^2)\} = 75 \text{ kN/m}^2 \qquad \text{(Equation 6.31)}$$

Assume a value of σ_z between γz and σ_{max} say 70 kN/m^2

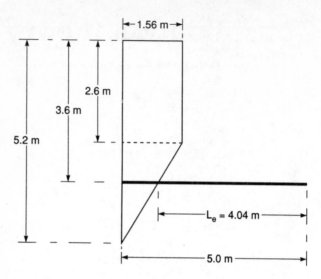

Fig. Q6.12

Tensile force in element:

$$T = 0.25 \times 70 \times 1.20 \times 0.65 = 13.7 \, \text{kN} \qquad \text{(Equation 6.30)}$$

Available frictional resistance:

$$R = 2 \times 0.065 \times 4.04 \times 70 \times \tan 35° = 25.7 \, \text{kN} \qquad \text{(Equation 6.32)}$$

Factor of safety against bond failure:

$$F_b = R/T = 25.7/13.7 = 1.9$$

Tensile stress $= (13.7 \times 10^3)/(65 \times 3) = 70 \, \text{kN/m}^2$
Factor of safety against tensile failure:

$$F_t = 340/70 = 4.8$$

7.1

Total change in thickness:

$$\Delta H = 7.82 - 6.02 = 1.80\,\text{mm}$$

Average thickness $= 15.30 + \dfrac{1.80}{2} = 16.20\,\text{mm}$

Length of drainage path:

$$d = \frac{16.20}{2} = 8.10\,\text{mm}$$

Root time plot (Fig. Q7.la):

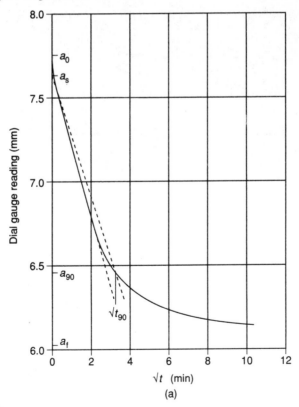

Fig. Q7.1(a)

$$\sqrt{t_{90}} = 3.3 \quad \therefore \quad t_{90} = 10.9 \text{ min}$$

$$c_v = \frac{0.848 d^2}{t_{90}} = \frac{0.848 \times 8.10^2}{10.9} \times \frac{1440 \times 365}{10^6} = 2.7 \text{ m}^2/\text{year}$$

$$r_0 = \frac{7.82 - 7.64}{7.82 - 6.02} = \frac{0.18}{1.80} = 0.100$$

$$r_p = \frac{10(7.64 - 6.45)}{9(7.82 - 6.02)} = \frac{10 \times 1.19}{9 \times 1.80} = 0.735$$

$$r_s = 1 - (0.100 + 0.735) = 0.165$$

Log time plot (Fig. Q7.1b):

$$t_{50} = 2.6 \text{ min}$$

$$c_v = \frac{0.196 d^2}{t_{50}} = \frac{0.196 \times 8.10^2}{2.6} \times \frac{1440 \times 365}{10^6} = 2.6 \text{ m}^2/\text{year}$$

$$r_0 = \frac{7.82 - 7.63}{7.82 - 6.02} = \frac{0.19}{1.80} = 0.106$$

$$r_p = \frac{7.63 - 6.23}{7.82 - 6.02} = \frac{1.40}{1.80} = 0.778$$

$$r_s = 1 - (0.106 + 0.778) = 0.116$$

Final void ratio:

$$e_1 = w_1 G_s = 0.232 \times 2.72 = 0.631$$

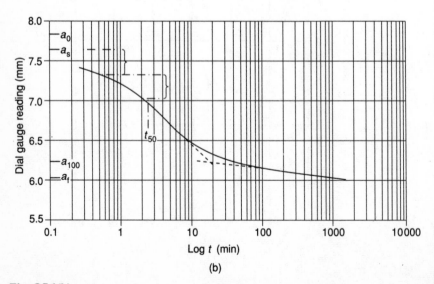

(b)

Fig. Q7.1(b)

$$\frac{\Delta e}{\Delta H} = \frac{1 + e_0}{H_0} = \frac{1 + e_1 + \Delta e}{H_0}$$

i.e.

$$\frac{\Delta e}{1.80} = \frac{1.631 + \Delta e}{17.10}$$

$$\therefore \quad \Delta e = \frac{2.936}{15.30} = 0.192$$

Initial void ratio:

$$e_0 = 0.631 + 0.192 = 0.823$$

Then:

$$m_v = \frac{1}{1 + e_0} \times \frac{e_0 - e_1}{\sigma_1' - \sigma_0'} = \frac{1}{1.823} \times \frac{0.192}{0.107} = 0.98 \ \text{m}^2/\text{MN}$$

$$k = c_v m_v \gamma_w = \frac{2.65 \times 0.98 \times 9.8}{60 \times 1440 \times 365 \times 10^3} = 8.1 \times 10^{-10} \ \text{m/s}$$

7.2

Using equation 7.7 (one-dimensional method):

$$s_c = \frac{e_0 - e_1}{1 + e_0} H$$

Appropriate values of e are obtained from Fig. Q7.2. The clay will be divided into four sub-layers, hence $H = 2000$ mm.

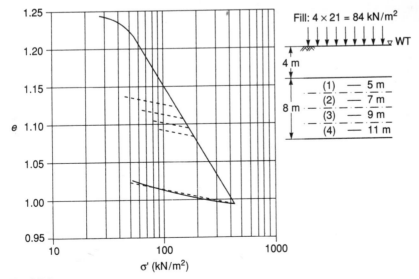

Fig. Q7.2

Settlement

Layer	σ_0' (kN/m^2)	σ_1' (kN/m^2)	e_0	e_1	$e_0 - e_1$	s_c (mm)
1	46.0*	130.0†	1.236	1.123	0.113	101
2	64.4	148.4	1.200	1.108	0.092	84
3	82.8	166.8	1.172	1.095	0.077	71
4	101.2	185.2	1.150	1.083	0.067	62
						318

*5×9.2
†$46.0 + 84$

Heave

Layer	σ_0' (kN/m^2)	σ_1' (kN/m^2)	e_0	e_1	$e_0 - e_1$	s_c (mm)
1	130.0	46.0	1.123	1.136	− 0.013	− 12
2	148.4	64.4	1.108	1.119	− 0.011	− 10
3	166.8	82.8	1.095	1.104	− 0.009	− 9
4	185.2	101.2	1.083	1.091	− 0.008	− 7
						−38

7.3

$$U = f(T_v) = f\left(\frac{c_v t}{d^2}\right)$$

Hence if c_v is constant:

$$\frac{t_1}{t_2} = \frac{d_1^2}{d_2^2}$$

where '1' refers to the oedometer specimen and '2' refers to the clay layer.
For open layers:

$$d_1 = 9.5 \text{ mm} \quad \text{and} \quad d_2 = 2500 \text{ mm}$$

$\therefore$ for $U = 0.50$, $\quad t_2 = t_1 \times \dfrac{d_2^2}{d_1^2}$

$$= \frac{20}{60 \times 24 \times 365} \times \frac{2500^2}{9.5^2} = 2.63 \text{ years}$$

for $U < 0.60$, $\quad T_v = \dfrac{\pi}{4} U^2$ (equation 7.24a)

$\therefore \quad t_{0.30} = t_{0.50} \times \dfrac{0.30^2}{0.50^2}$

$$= 2.63 \times 0.36 = 0.95 \text{ years}$$

7.4

The layer is open,

$$\therefore \quad d = \frac{8}{2} = 4 \text{ m}$$

$$T_v = \frac{c_v t}{d^2} = \frac{2.4 \times 3}{4^2} = 0.450$$

$$u_i = \Delta\sigma = 84 \text{ kN/m}^2$$

The excess pore water pressure is given by equation 7.21:

$$u_e = \sum_{m=0}^{m=\infty} \frac{2u_i}{M} \left(\sin \frac{Mz}{d} \right) \exp(-M^2 T_v)$$

In this case, $z = d$

$$\therefore \quad \sin \frac{Mz}{d} = \sin M$$

where

$$M = \frac{\pi}{2}, \frac{3\pi}{2}, \frac{5\pi}{2} \ldots$$

M	$\sin M$	$M^2 T_v$	$\exp(-M^2 T_v)$
$\frac{\pi}{2}$	$+1$	1.110	0.329
$\frac{3\pi}{2}$	-1	9.993	4.57×10^{-5}

$$\therefore \quad u_e = 2 \times 84 \times \frac{2}{\pi} \times 1 \times 0.329 \quad \text{(other terms negligible)}$$

$$= 35.2 \text{ kN/m}^2$$

7.5

The layer is open,

$$\therefore \quad d = \frac{6}{2} = 3 \text{ m}$$

$$T_v = \frac{c_v t}{d^2} = \frac{1.0 \times 3}{3^2} = 0.333$$

The layer thickness will be divided into six equal parts, i.e. $m = 6$.
  For an open layer:

$$T_v = 4 \frac{n}{m^2} \beta$$

$$\therefore \quad n\beta = \frac{0.333 \times 6^2}{4} = 3.00$$

The value of n will be taken as 12 (i.e. $\Delta t = 3/12 = 1/4$ year), making $\beta = 0.25$. The computation is set out below, all pressures having been multiplied by 10:

$$u_{i,j+1} = u_{i,j} + 0.25(u_{i-1,j} + u_{i+1,j} - 2u_{i,j})$$

$i \backslash j$	0	1	2	3	4	5	6	7	8	9	10	11	12
0	0	0	0	0	0	0	0	0	0	0	0	0	0
1	500	350	275	228	195	171	151	136	123	112	102	94	87
2	400	400	362	325	292	264	240	219	201	185	171	158	146
3	300	300	300	292	277	261	245	230	215	201	187	175	163
4	200	200	200	200	198	193	189	180	171	163	154	145	137
5	100	100	100	100	100	99.5	98	96	93	89	85	81	77
6	0	0	0	0	0	0	0	0	0	0	0	0	0

The initial and three-year isochrones are plotted in Fig. Q7.5.

Area under initial isochrone = 180 units

Area under three-year isochrone = 63 units

The average degree of consolidation is given by equation 7.25. Thus:

$$U = 1 - \frac{63}{180} = 0.65$$

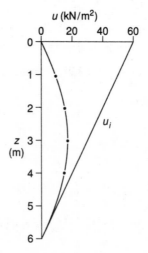

Fig. Q7.5

7.6

At the top of the clay layer the decrease in pore water pressure is $4\gamma_w$. At the bottom of the clay layer the pore water pressure remains constant. Hence at the centre of the clay layer,

$$\Delta\sigma' = 2\gamma_w = 2 \times 9.8 = 19.6\,\text{kN/m}^2$$

The final consolidation settlement (one-dimensional method) is:

$$s_c = m_v\Delta\sigma'H = 0.83 \times 19.6 \times 8 = 130\,\text{mm}$$

Corrected time:

$$t = 2 - \frac{1}{2}\left(\frac{40}{52}\right) = 1.615 \text{ years}$$

$$\therefore\quad T_v = \frac{c_vt}{d^2} = \frac{4.4 \times 1.615}{4^2} = 0.444$$

From Fig. 7.18 (Curve 1), $U = 0.73$

Settlement after two years $= Us_c = 0.73 \times 130 = 95$ mm

7.7

The clay layer is thin relative to the dimensions of the raft, and therefore the one-dimensional method is appropriate. The clay layer can be considered as a whole. See Fig. Q7.7.

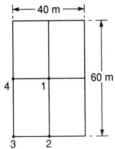

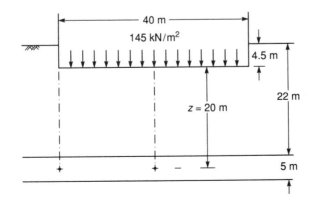

Fig. Q7.7

Point	m	n	I_r	$\Delta\sigma$ (kN/m²)	s_c^* (mm)
1	$30/20 = 1.5$	$20/20 = 1.0$	0.194 ($\times 4$)	113	124
2	$60/20 = 3.0$	$20/20 = 1.0$	0.204 ($\times 2$)	59	65
3	$60/20 = 3.0$	$40/20 = 2.0$	0.238 ($\times 1$)	35	38
4	$30/20 = 1.5$	$40/20 = 2.0$	0.224 ($\times 2$)	65	72

$$^*s_c = m_v \Delta\sigma' H = 0.22 \times \Delta\sigma' \times 5 = 1.1\Delta\sigma' \quad \text{(mm)} \quad (\Delta\sigma' = \Delta\sigma)$$

7.8

Due to the thickness of the clay layer relative to the size of the foundation, there will be significant lateral strain in the clay and the Skempton–Bjerrum method is appropriate. The clay is divided into six sub-layers (Fig. Q7.8) for the calculation of consolidation settlement.

(a) Immediate settlement:

$$H/B = \frac{30}{35} = 0.86$$

$$D/B = \frac{2}{35} = 0.06$$

From Fig. 5.15 (circle), $\mu_1 = 0.32$ and $\mu_0 = 1.0$:

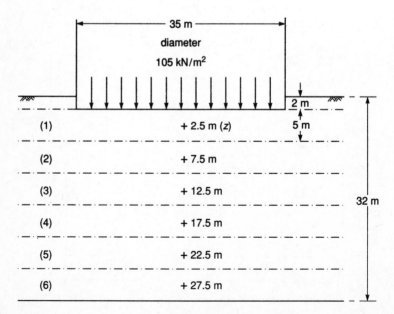

Fig. Q7.8

$$s_i = \mu_0 \mu_1 \frac{qB}{E_u} = 1.0 \times 0.32 \times \frac{105 \times 35}{40} = 30 \text{ mm}$$

(b) Consolidation settlement:

Layer	z (m)	D/z	I_c^*	$\Delta\sigma$ (kN/m^2)	$s_{od}^\dagger$ (mm)
1	2.5	14	0.997	107	73.5
2	7.5	4.67	0.930	98	68.6
3	12.5	2.80	0.804	84	58.8
4	17.5	2.00	0.647	68	47.6
5	22.5	1.55	0.505	53	37.1
6	27.5	1.27	0.396	42	29.4
					315.0

* From Fig. 5.9
† $s_{od} = m_v \Delta\sigma' H = 0.14 \times \Delta\sigma' \times 5 = 0.70 \Delta\sigma'$ $(\Delta\sigma' = \Delta\sigma)$

Now:

$$H/B = \frac{30}{35} = 0.86 \quad \text{and} \quad A = 0.65$$

$$\therefore \text{ from Fig. 7.12,} \quad \mu = 0.79$$

$$\therefore \quad s_c = \mu s_{od} = 0.79 \times 315 = 250 \text{ mm}$$

Total settlement:

$$s = s_i + s_c$$

$$= 30 + 250 = 280 \text{ mm}$$

7.9

Without sand drains:

$$U_v = 0.25 \quad \therefore \quad T_v = 0.049 \quad \text{(from Fig. 7.18)}$$

$$\therefore \quad t = \frac{T_v d^2}{c_v} = \frac{0.049 \times 8^2}{c_v}$$

With sand drains:

$$R = 0.564S = 0.564 \times 3 = 1.69 \text{ m}$$

$$n = \frac{R}{r} = \frac{1.69}{0.15} = 11.3$$

$$T_r = \frac{c_h t}{4R^2} = \frac{c_h}{4 \times 1.69^2} \times \frac{0.049 \times 8^2}{c_v} \quad \text{(and } c_h = c_v\text{)}$$

$$= 0.275$$

$$\therefore \quad U_r = 0.73 \quad \text{(from Fig. 7.30)}$$

Using equation 7.40:

$$(1 - U) = (1 - U_v)(1 - U_r)$$

$$= (1 - 0.25)(1 - 0.73) = 0.20$$

$$\therefore \quad U = 0.80$$

7.10

Without sand drains:

$$U_v = 0.90 \quad \therefore \quad T_v = 0.848$$

$$\therefore \quad t = \frac{T_v d^2}{c_v} = \frac{0.848 \times 10^2}{9.6} = 8.8 \text{ years}$$

With sand drains:

$$R = 0.564S = 0.564 \times 4 = 2.26 \text{ m}$$

$$n = \frac{R}{r} = \frac{2.26}{0.15} = 15$$

$$\frac{T_r}{T_v} = \frac{c_h}{c_v} \frac{d^2}{4R^2} \quad \text{(same } t\text{)}$$

$$\therefore \quad \frac{T_r}{T_v} = \frac{14.0}{9.6} \times \frac{10^2}{4 \times 2.26^2} = 7.14 \quad (1)$$

Using equation 7.40:

$$(1 - U) = (1 - U_v)(1 - U_r)$$

$$\therefore \quad (1 - 0.90) = (1 - U_v)(1 - U_r)$$

$$\therefore \quad (1 - U_v)(1 - U_r) = 0.10 \quad (2)$$

An iterative solution is required using (1) and (2) above, an initial value of U_v being estimated.

U_v	T_v	$T_r = 7.14 T_v$	U_r	$(1 - U_v)(1 - U_r)$
0.40	0.1256	0.897	0.97	$0.60 \times 0.03 = 0.018$
0.30	0.0707	0.505	0.87	$0.70 \times 0.13 = 0.091$
0.29	0.0660	0.471	0.85	$0.71 \times 0.15 = 0.107$
0.295	0.0683	0.488	0.86	$0.705 \times 0.14 = 0.099$

Thus:

$$U_v = 0.295 \quad \text{and} \quad U_r = 0.86$$

$$\therefore \quad t = 8.8 \times \frac{0.0683}{0.848} = 0.7 \text{ years}$$

Bearing capacity

8.1

(a) The ultimate bearing capacity is given by equation 8.3:

$$q_f = \tfrac{1}{2}\gamma B N_\gamma + c N_c + \gamma D N_q$$

For $\varphi_u = 0$:

$$N_\gamma = 0, \quad N_c = 5.14, \quad N_q = 1$$

$$\therefore \quad q_f = (105 \times 5.14) + (21 \times 1 \times 1) = 540 + 21 \text{ kN/m}^2$$

The net ultimate bearing capacity is:

$$q_{nf} = q_f - \gamma D = 540 \text{ kN/m}^2$$

The net foundation pressure is:

$$q_n = q - \gamma D = \frac{425}{2} - (21 \times 1) = 192 \text{ kN/m}^2$$

The factor of safety (equation 8.7) is:

$$F = \frac{q_{nf}}{q_n} = \frac{540}{192} = 2.8$$

(b) For $\varphi' = 28°$:

$$N_\gamma = 13, \quad N_c = 26, \quad N_q = 15 \qquad \text{(Fig. 8.4)}$$

$$\gamma' = 21 - 9.8 = 11.2 \text{ kN/m}^3$$

$$\therefore \quad q_f = (\tfrac{1}{2} \times 11.2 \times 2 \times 13) + (10 \times 26) + (11.2 \times 1 \times 15)$$

$$= 146 + 260 + 168 = 574 \text{ kN/m}^2$$

$$q_{nf} = 574 - 11.2 = 563 \text{ kN/m}^2$$

$$F = \frac{563}{192} = 2.9$$

($q_n = 192 \text{ kN/m}^2$ assumes that backfilled soil on the footing slab is included in the load of 425 kN/m.)

8.2 $c' = 0$

(a) For $\varphi' = 38°$:

$$N_\gamma = 67, \quad N_q = 49$$

$$\therefore \quad q_{nf} = \tfrac{1}{2}\gamma B N_\gamma + \gamma D(N_q - 1) \qquad \text{(from equation 8.3)}$$

$$= (\tfrac{1}{2} \times 18 \times 1.5 \times 67) + (18 \times 0.75 \times 48)$$

$$= 905 + 648 = 1553 \text{ kN/m}^2$$

$$q_n = \frac{500}{1.5} - (18 \times 0.75) = 320 \text{ kN/m}^2$$

$$\therefore \quad F = \frac{q_{nf}}{q_n} = \frac{1553}{320} = 4.8$$

(b) Inclination factors (equations 8.10):

$$i_\gamma = \left(1 - \frac{\alpha}{\varphi}\right)^2 = \left(1 - \frac{10}{38}\right)^2 = 0.54$$

$$i_q = \left(1 - \frac{\alpha}{90}\right)^2 = \left(1 - \frac{10}{90}\right)^2 = 0.79$$

$$\therefore \quad q_{nf} = (905 \times 0.54) + (648 \times 0.79) = 489 + 512 = 1001 \text{ kN/m}^2$$

$$\therefore \quad F = \frac{1001}{320} = 3.1$$

8.3 $A = 4.50 m \times 2.25 L$; $Df = 3.50 m$

$FS = 3$

$$D/B = \frac{3.50}{2.25} = 1.55$$

From Fig. 8.5, for a square foundation:

$$N_c = 8.1$$

For a rectangular foundation ($L = 4.50$ m; $B = 2.25$ m):

$$N_c = \left(0.84 + 0.16\frac{B}{L}\right)8.1 = 7.45$$

Using equation 8.8:

$$q_{nf} = q_f - \gamma D = c_u N_c = 135 \times 7.45 = 1006 \text{ kN/m}^2$$

For $F = 3$:

$$q_n = \frac{1006}{3} = 335 \text{ kN/m}^2$$

$$\therefore \quad q = q_n + \gamma D = 335 + (20 \times 3.50) = 405 \text{ kN/m}^2$$

$$\therefore \quad \text{Allowable load} = 405 \times 4.50 \times 2.25 = 4100 \text{ kN}$$

8.4

For $\varphi' = 40°$:

$$N_\gamma = 95, \quad N_q = 64$$

$$q_{nf} = 0.4\gamma B N_\gamma + \gamma D(N_q - 1) \qquad \text{(from equation 8.4)}$$

(a) Water table 5 m below ground level:

$$q_{nf} = (0.4 \times 17 \times 2.5 \times 95) + (17 \times 1 \times 63)$$

$$= 1615 + 1071 = 2686 \text{ kN/m}^2$$

$$q_n = 400 - 17 = 383 \text{ kN/m}^2$$

$$F = \frac{2686}{383} = 7.0$$

(b) Water table 1 m below ground level (i.e. at foundation level):

$$\gamma' = 20 - 9.8 = 10.2 \text{ kN/m}^3$$

$$q_{nf} = (0.4 \times 10.2 \times 2.5 \times 95) + (17 \times 1 \times 63)$$

$$= 969 + 1071 = 2040 \text{ kN/m}^2$$

$$F = \frac{2040}{383} = 5.3$$

(c) Water table at ground level with upward hydraulic gradient 0.2:

$$(\gamma' - j) = 10.2 - (0.2 \times 9.8) = 8.2 \text{ kN/m}^3$$

$$q_{nf} = (0.4 \times 8.2 \times 2.5 \times 95) + (8.2 \times 1 \times 63)$$

$$= 779 + 517 = 1296 \text{ kN/m}^2$$

$$F = \frac{1296}{383} = 3.4$$

8.5

Undrained shear, for $\varphi_u = 0$:

$$N_\gamma = 0, \quad N_c = 5.14, \quad N_q = 1$$

$$q_{nf} = 1.2 c_u N_c \qquad \text{(from equation 8.4)}$$

$$= 1.2 \times 100 \times 5.14 = 617 \text{ kN/m}^2$$

$$q_n = \frac{q_{nf}}{F} = \frac{617}{3} = 206 \text{ kN/m}^2$$

$$q = q_n + \gamma D = 206 + 21 = 227 \text{ kN/m}^2$$

Drained shear, for $\varphi' = 27°$:

$$N_\gamma = 11, \quad N_c = 24, \quad N_q = 13$$

$$\begin{aligned} q_{nf} &= 0.4\gamma' B N_\gamma + 1.2c' N_c + \gamma' D(N_q - 1) \\ &= (0.4 \times 11.2 \times 4 \times 11) + (1.2 \times 15 \times 24) + (11.2 \times 1 \times 12) \\ &= 197 + 432 + 134 = 763 \text{ kN/m}^2 \end{aligned}$$

$$q = \frac{763}{3} + 21 = 275 \text{ kN/m}^2$$

Consolidation settlement; the clay will be divided into three sub-layers (Fig. Q8.5):

Layer	z (m)	m, n	I_r	$\Delta\sigma'$ (kN/m^2)	s_{od} (mm)
1	2	1.00	0.175	$0.700q_n$	$0.182q_n$
2	6	0.33	0.044	$0.176q_n$	$0.046q_n$
3	10	0.20	0.017	$0.068q_n$	$0.018q_n$
					$0.246q_n$

Diameter of equivalent circle: $B = 4.5$ m

$$H/B = \frac{12}{4.5} = 2.7 \quad \text{and} \quad A = 0.42 \quad \text{(given)}$$

$$\therefore \quad \mu = 0.60 \quad \text{(Fig. 7.12)}$$

$$s_c = 0.60 \times 0.246q_n = 0.147q_n \quad \text{(mm)}$$

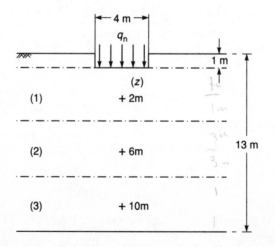

Fig. Q8.5

For $s_c = 30$ mm:

$$q_n = \frac{30}{0.147} = 204 \text{ kN/m}^2$$

$$q = 204 + 21 = 225 \text{ kN/m}^2$$

Thus the allowable bearing capacity is 225 kN/m², settlement being the limiting criterion.

8.6

$$D/B = \frac{8}{4} = 2.0$$

From Fig. 8.5, for a strip, $N_c = 7.1$. Then from equation 8.12:

$$F = \frac{c_u N_c}{\gamma D} = \frac{40 \times 7.1}{20 \times 8} = 1.8$$

8.7

Depth (m)	N	σ'_v (kN/m²)*	C_N	N_1
0.70	6	—	—	—
1.35	9	23	1.90	17
2.20	10	37	1.55	15
2.95	8	50	1.37	11
3.65	12	58	1.28	15
4.40	13	65	1.23	16
5.15	17	—	—	—
6.00	23	—	—	—

* Using $\gamma = 17 \text{ kN/m}^3$ and $\gamma' = 10 \text{ kN/m}^3$.

(a) *Terzaghi and Peck.* Use N_1 values between depths of 1.2 m and 4.7 m, the average value being 15. For $B = 3.5$ m and $N = 15$ the provisional value of bearing capacity, using Fig. 8.10, is 150 kN/m². The water table correction factor (equation 8.16) is:

$$C_w = 0.5 + 0.5(3.0/4.7) = 0.82$$

Thus:

$$q_a = 150 \times 0.82 = 120 \text{ kN/m}^2$$

(b) *Meyerhof.* Use uncorrected N values between depths of 1.2 m and 4.7 m, the average value being 10. For $B = 3.5$ m and $N = 10$ the provisional value of bearing capacity, using Fig. 8.10, is 90 kN/m². This value is increased by 50%.

Thus:

$$q_a = 90 \times 1.5 = 135 \text{ kN/m}^2$$

(c) *Burland and Burbidge.* Using Fig. 8.12, for $B = 3.5$ m, $z_1 = 2.5$ m. Use N values between depths of 1.2 m and 3.7 m, the average value being 10. From equation 8.18:

$$I_c = 1.71/10^{1.4} = 0.068$$

From equation 8.19a, with $s = 25$ mm:

$$q = 25/(3.5^{0.7} \times 0.068) = 150 \text{ kN/m}^2$$

8.8

(a) *Buisman–DeBeer*: refer to Fig. Q8.8.

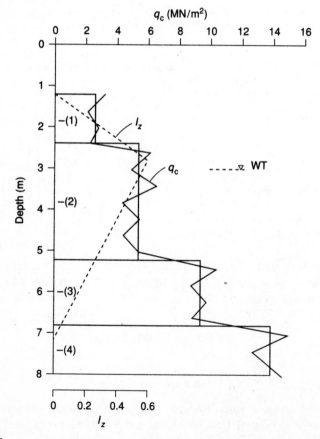

Fig. Q8.8

Layer	Δz (m)	q_c (MN/m^2)	z_c (m)	σ_0' (kN/m^2)	$\Delta\sigma$ (kN/m^2)	$\dfrac{\sigma_0'+\Delta\sigma}{\sigma_0'}$	log	$1.53\dfrac{\sigma_0'\Delta z}{q_c}$	Δs (mm)
1	1.2	2.6	1.8	29	124	5.28	0.723	20.5	14.8
2	2.8	5.4	3.8	55	58	2.05	0.312	43.6	13.6
3	1.6	9.5	6.0	76	23	1.30	0.114	19.6	2.2
4	1.2	14.0	7.4	88	15	1.17	0.068	11.5	0.8
									$s = 31.4$

(b) *Schmertmann*

Layer	Δz (m)	q_c (MN/m^2)	$E = 2q_c$ (MN/m^2)	I_z	$\dfrac{I_z}{E}\Delta z$ (mm^3/MN)
1	1.2	2.6	5.2	0.24	0.0554
2	2.8	5.4	10.8	0.45	0.1167
3	1.6	9.5	19.0	0.16	0.0135
4	1.2	14.0	28.0	—	—
					0.1856

$$C_1 = 1 - 0.5\frac{\sigma_0'}{q_n} = 1 - \frac{0.5 \times 1.2 \times 16}{130} = 0.93$$

$$C_2 = 1 \quad \text{(say)}$$

$$\therefore \quad s = C_1 C_2 q_n \sum \frac{I_z}{E}\Delta z = 0.93 \times 1 \times 130 \times 0.1856 = 22 \text{ mm}$$

8.9

At pile base level:

$$c_u = 220 \text{ kN/m}^2$$

$$\therefore \quad q_f = c_u N_c = 220 \times 9 = 1980 \text{ kN/m}^2$$

Disregard skin friction over a length of $2B$ above the under-ream. Between 4 m and 17.9 m:

$$\overline{\sigma}_0' = 10.95\gamma' = 10.95 \times 11.2 = 122.6 \text{ kN/m}^2$$

$$\therefore \quad f_s = \beta\overline{\sigma}_0' = 0.7 \times 122.6 = 86 \text{ kN/m}^2$$

Then:

$$Q_f = A_b q_f + A_s f_s$$

$$= \left(\frac{\pi}{4} \times 3^2 \times 1980\right) + (\pi \times 1.05 \times 13.9 \times 86)$$

$$= 13\,996 + 3941 = 17\,937 \text{ kN}$$

Allowable load:

$$\text{(a)} \quad \frac{Q_f}{2} = \frac{17\,937}{2} = 8968 \text{ kN}$$

$$\text{(b)} \quad \frac{A_b q_f}{3} + A_s f_s = \frac{13\,996}{3} + 3941 = 8606 \text{ kN}$$

i.e. allowable load = 8600 kN.

Adding $\frac{1}{3}(\gamma D A_b - W)$, the allowable load becomes 9200 kN.

8.10

$$q_f = 9c_u = 9 \times 145 = 1305 \text{ kN/m}^2$$

$$f_s = \alpha c_u = 0.40 \times 105 = 42 \text{ kN/m}^2$$

For a single pile:

$$Q_f = A_b q_f + A_s f_s$$
$$= (\pi/4 \times 0.6^2 \times 1305) + (\pi \times 0.6 \times 15 \times 42)$$
$$= 369 + 1187 = 1556 \text{ kN}$$

Assuming single pile failure and a group efficiency of 1, the ultimate load on the pile group is $(1556 \times 36) = 56\,016$ kN. The width of the group is 12.6 m, and hence the ultimate load, assuming block failure and taking the full undrained strength on the perimeter, is given by:

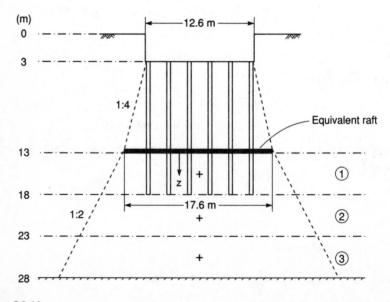

Fig. Q8.10

$$(12.6^2 \times 1305) + (4 \times 12.6 \times 15 \times 105)$$

$$= 207\,180 + 79\,380 = 286\,560\,\text{kN}$$

(Even if the remoulded strength were used there would be no likelihood of block failure.) Thus the load factor is $(56\,016/21\,000) = 2.7$.

Settlement is estimated using the equivalent raft concept. The equivalent raft is located 10 m ($\frac{2}{3} \times 15$ m) below the top of the piles and is 17.6 m wide (see Fig. Q8.10). Assume that the load on the equivalent raft is spread at 2 : 1 to the underlying clay. Thus the pressure on the equivalent raft is:

$$q = 21\,000/17.6^2 = 68\,\text{kN/m}^2$$

Immediate settlement:

$$H/B = 15/17.6 = 0.85$$

$$D/B = 13/17.6 = 0.74$$

$$L/B = 1$$

Hence from Fig. 5.15:

$$\mu_0 = 0.78 \quad \text{and} \quad \mu_1 = 0.41$$

Thus, using Equation 5.28:

$$s_i = 0.78 \times 0.41 \times 68 \times 17.6/65 = 6\,\text{mm}$$

Consolidation settlement:

Layer	z (m)	Area (m^2)	$\Delta\sigma$ (kN/m^2)	$m_v \Delta \sigma H$ (mm)
1	2.5	20.1^2	52.0	20.8
2	7.5	25.1^2	33.3	13.3
3	12.5	30.1^2	23.2	9.3
				43.4 (s_{od})

Equivalent diameter $= 19.86$ m: thus $H/B = 15/19.86 = 0.76$. Now $A = 0.28$, hence from Fig. 7.12, $\mu = 0.56$. Then, from equation 7.10:

$$s_c = 0.56 \times 43.4 = 24\,\text{mm}$$

The total settlement is $(6 + 24) = 30$ mm.

8.11

At base level, $N = 26$. Then using equation 8.35:

$$q_f = 40N \frac{D_b}{B} = 40 \times 26 \times \frac{2}{0.25} = 8320\,\text{kN/m}^2$$

(Check: $< 400N$, i.e. $400 \times 26 = 10\,400\,\text{kN/m}^2$)

Over the length embedded in sand:

$$\overline{N} = 21 \qquad \left(\text{i.e. } \frac{18 + 24}{2}\right)$$

Using equation 8.36:

$$f_s = 2\overline{N} = 2 \times 21 = 42\,\text{kN/m}^2$$

For a single pile:

$$Q_f = A_b q_f + A_s f_s$$
$$= (0.25^2 \times 8320) + (4 \times 0.25 \times 2 \times 42)$$
$$= 520 + 84 = 604\,\text{kN}$$

For the pile group, assuming a group efficiency of 1.2:

$$\Sigma Q_f = 1.2 \times 9 \times 604 = 6523\,\text{kN}$$

For a load factor of 2.5, the allowable load is:

$$\frac{\Sigma Q_f}{2.5} = \frac{6523}{2.5} = 2609\,\text{kN}$$

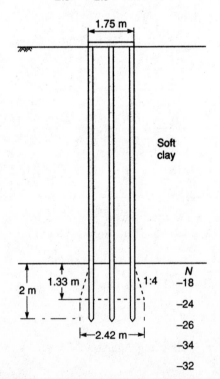

Fig. Q8.11

Referring to Fig. Q8.11, the equivalent raft is 2.42 m square. The average value of N over depth B below the equivalent raft is:

$$\overline{N} = (24 + 26 + 34 + 32)/4 = 29$$

From Fig. 8.10, for $B = 2.42$ m and $N = 29$, the value of q_a is 325 kN/m². Then, if the settlement is not to exceed 25 mm, the allowable load is:

$$\Sigma Q = 325 \times 2.42^2 = 1900 \, \text{kN}$$

Thus, the allowable load on the pile group is 1900 kN, settlement being the limiting criterion.

8.12

Using equation 8.45:

$$Q_f = \pi DL\alpha c_u + \frac{\pi}{4}(D^2 - d^2)c_u N_c$$

$$= (\pi \times 0.2 \times 5 \times 0.6 \times 110) + \frac{\pi}{4}(0.2^2 - 0.1^2)110 \times 9$$

$$= 207 + 23 = 230 \, \text{kN}$$

Stability of Slopes

9.1

Referring to Fig. Q9.1:

$$W = 41.7 \times 19 = 792 \text{ kN/m}$$

$$\text{Arc length AB} = \frac{\pi}{180} \times 73 \times 9.0 = 11.5 \text{ m}$$

$$\text{Arc length BC} = \frac{\pi}{180} \times 28 \times 9.0 = 4.4 \text{ m}$$

The factor of safety is given by:

$$F = \frac{r\,\Sigma(c_u L_a)}{Wd} = \frac{9.0\{(20 \times 4.4) + (35 \times 11.5)\}}{792 \times 3.9} = 1.43$$

Depth of tension crack:

$$z_0 = \frac{2c_u}{\gamma} = \frac{2 \times 20}{19} = 2.1 \text{ m}$$

$$\text{Arc length BD} = \frac{\pi}{180} \times 13\tfrac{1}{2} \times 9.0 = 2.1 \text{ m}$$

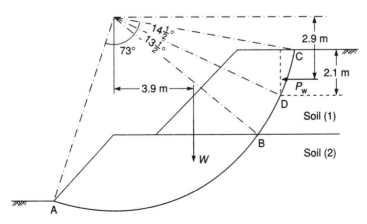

Fig. Q9.1

$$P_w = \tfrac{1}{2} \times 9.8 \times 2.1^2 = 21.6 \, kN/m$$

$$F = \frac{9.0\,\{(20 \times 2.1) + (35 \times 11.5)\}}{(792 \times 3.9) + (21.6 \times 2.9)} = 1.27$$

9.2

$F = r \Sigma c_u l_a$ — design restsifty moment

w_d — design disturbing moment

$$\varphi_u = 0$$

Depth factor:

$$D = \frac{11}{9} = 1.22$$

Using equation 9.2 with $F = 1.0$:

$$N_s = \frac{c_u}{F\gamma H} = \frac{30}{1.0 \times 19 \times 9} = 0.175$$

Hence from Fig. 9.3:

$$\beta = 50°$$

For $F = 1.2$:

$$N_s = \frac{30}{1.2 \times 19 \times 9} = 0.146 \quad \therefore \quad \beta = 27°$$

9.3

Refer to Fig. Q9.3:

Slice no.	$h \cos \alpha$ (m)	$h \sin \alpha$ (m)	u/γ_w (m)	u (kN/m²)	l (m)	ul (kN/m)
1	0.5	—	0.7	6.9	1.2	8
2	1.2	− 0.1	1.7	16.7	2.0	33
3	2.4	—	3.0	29.4	2.0	59
4	3.4	0.2	3.9	38.2	2.0	76
5	4.3	0.5	4.7	46.1	2.1	97
6	4.9	0.9	5.1	50.0	2.1	105
7	5.3	1.4	5.7	55.9	2.1	117
8	5.7	1.8	5.8	56.8	2.1	119
9	5.9	2.4	5.9	57.8	2.2	127
10	5.9	2.9	6.0	58.8	2.2	129
11	5.6	3.3	5.7	55.9	2.3	129
12	5.2	3.5	5.2	51.0	2.4	122
13	4.6	3.7	4.5	44.1	2.5	110
14	3.4	3.2	3.4	33.3	2.7	90
15	1.6	1.9	1.8	17.6	2.9	51
	59.9	25.6			32.8	1372
					(check)	

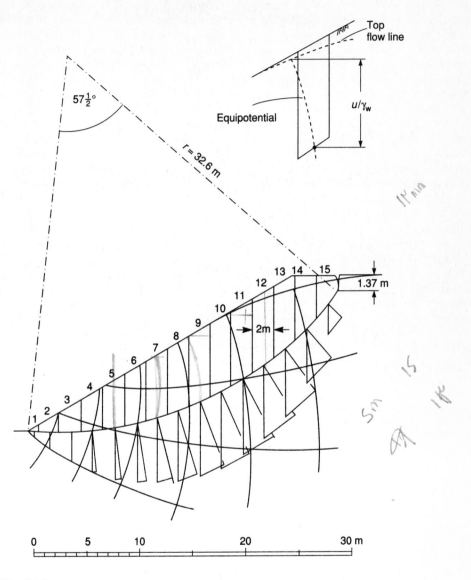

Fig. Q9.3

$$\Sigma W \cos \alpha = \gamma b \, \Sigma h \cos \alpha = 21 \times 2 \times 59.9 = 2516 \, \text{kN/m}$$

$$\Sigma W \sin \alpha = \gamma b \, \Sigma h \sin \alpha = 21 \times 2 \times 25.6 = 1075 \, \text{kN/m}$$

$$\Sigma (W \cos \alpha - ul) = 2516 - 1372 = 1144 \, \text{kN/m}$$

$$\text{Arc length } L_a = \frac{\pi}{180} \times 57\tfrac{1}{2} \times 32.6 = 32.7 \, \text{m}$$

The factor of safety is given by:

$$F = \frac{c'L_a + \tan \varphi' \, \Sigma(W \cos \alpha - ul)}{\Sigma W \sin \alpha}$$

$$= \frac{(8 \times 32.7) + (\tan 32° \times 1144)}{1075}$$

$$= 0.91$$

9.4

$$F = \frac{1}{\Sigma W \sin \alpha} \Sigma \left[\{c'b + (W - ub) \tan \varphi'\} \frac{\sec \alpha}{1 + (\tan \alpha \tan \varphi'/F)} \right]$$

$$c' = 8 \text{ kN/m}^2$$

$$\varphi' = 32°$$

$$c'b = 8 \times 2 = 16 \text{ kN/m}$$

$$W = \gamma bh = 21 \times 2 \times h = 42h \text{ kN/m}$$

$$\text{Try } F = 1.00$$

$$\tan \varphi'/F = 0.625$$

Values of u are as obtained in Q9.3.

Slice no.	h (m)	$W = \gamma bh$ (kN/m)	$\alpha°$	$W \sin \alpha$ (kN/m)	ub (kN/m)	$c'b + (W - ub)$ $\times \tan \varphi'$ (kN/m)	$\dfrac{\sec \alpha}{1 + (\tan \alpha \tan \varphi')/F}$	Product (kN/m)
1	0.5	21	−6	−2	8	24	1.078	26
2	1.3	55	−3½	−3	33	30	1.042	31
3	2.4	101	0	0	59	42	1.000	42
4	3.4	143	4	10	76	58	0.960	56
5	4.3	181	7½	24	92	72	0.931	67
6	5.0	210	11	40	100	85	0.907	77
7	5.5	231	14½	58	112	90	0.889	80
8	6.0	252	18½	80	114	102	0.874	89
9	6.3	265	22	99	116	109	0.861	94
10	6.5	273	26	120	118	113	0.854	97
11	6.5	273	30	136	112	117	0.850	99
12	6.3	265	34	148	102	118	0.847	100
13	5.9	248	38½	154	88	116	0.853	99
14	4.6	193	43	132	67	95	0.862	82
15	2.5	105	48	78	35	59	0.882	52
				1074				1091

$$F = \frac{1091}{1074} = 1.02 \quad \text{(assumed value 1.00)}$$

Thus:

$$F = 1.01$$

9.5

$$F = \frac{1}{\Sigma W \sin \alpha} \Sigma \left[\{c'b + W(1 - r_u) \tan \varphi'\} \frac{\sec \alpha}{1 + (\tan \alpha \tan \varphi')/F} \right]$$

$$c' = 16 \text{ kN/m}^2$$

$$\varphi' = 32°$$

$$r_u = 0.45$$

$$c'b = 16 \times 5 = 80 \text{ kN/m}$$

$$W = \gamma bh = 20 \times 5 \times h = 100h \text{ kN/m}$$

$$(1 - r_u) \tan \varphi' = 0.55 \tan 32° = 0.344$$

$$\text{Try } F = 1.20$$

$$\tan \varphi'/F = \tan 32°/1.20 = 0.521$$

Referring to Fig. Q9.5:

Slice no.	h (m)	$W = \gamma bh$ (kN/m)	$\alpha°$	$W \sin \alpha$ (kN/m)	$c'b + W(1 - r_u)$ $\times \tan \varphi'$ (kN/m)	$\dfrac{\sec \alpha}{1 + (\tan \alpha \tan \varphi')/F}$	Product (kN/m)
1	1.5	75	4	5	106	0.967	102
2	3.1	310	9	48	187	0.935	175
3	4.5	450	$15\frac{1}{2}$	120	235	0.907	213
4	5.3	530	21	190	262	0.893	234
5	6.0	600	28	282	286	0.887	254
6	5.0	500	35	287	252	0.894	225
7	3.4	340	43	232	197	0.920	181
8	1.4	28	49	21	90	0.953	86
				1185			1470

$$F = \frac{1470}{1185} = 1.24 \quad \text{(assumed value 1.20)}$$

A further trial yields:

$$F = 1.22$$

Write a computer program and confirm this result!

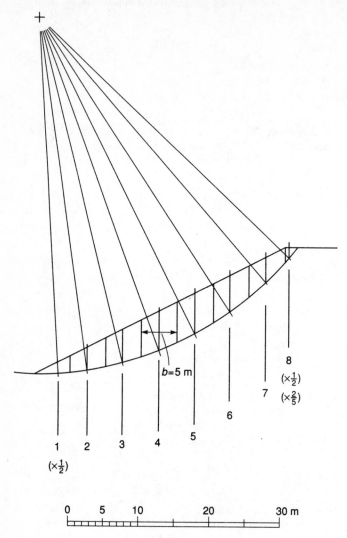

Fig. Q9.5

9.6

Water table at surface; the factor of safety is given by equation 9.12:

$$F = \frac{\gamma'}{\gamma_{sat}} \frac{\tan \varphi'}{\tan \beta}$$

i.e. $1.5 = \dfrac{9.2}{19} \dfrac{\tan 36°}{\tan \beta}$

$$\therefore \quad \tan \beta = 0.234$$

$$\beta = 13°$$

Water table well below surface; the factor of safety is given by equation 9.11:

$$F = \frac{\tan \varphi'}{\tan \beta}$$

$$= \frac{\tan 36°}{\tan 13°}$$

$$= 3.1$$